WHITHER COMETH HUMANKIND?

The Origins of Man.

Genesis and Science Agree!

CHARLES S. BROWN

WHITHER COMETH HUMANKIND?

The Origins of Man.

Genesis and Science Agree!

Charles S. Brown

http://www.crystalbooks.org

This Edition published in **New Zealand** by
Crystal Publishing.
P.O. Box 60042, Titirangi, Waitakere City,
Auckland, **New Zealand**

First published 2005
Second Edition 2007
Third Edition 2012

ISBN 978-0-9582813-2-4

Contents

Acknowledgements

For the specific subject matter of this Booklet, **Crystal Publishing** singularly and gratefully acknowledges **Ferrar Fenton** – long deceased from the physical world – yet whose monumental work of re-translating **The Bible** finally permitted certain key questions in centuries-old conflict between science and religion – specifically Creation versus Evolution – to be perfectly reconciled. Through his intuitively-correct translation of **The Book of Genesis**, particularly Chapters 1 and 2, he has singularly rendered *every other* "Genesis" translation – that does not accord with his correct, and thus powerfully-guided, insights – irrelevant. Near-future events will unequivocally bear out the truth of this statement.

Deriving from **Immutable Divine Law**, his crucial *spiritual* insights have therein returned to **The Creator** that which is **His**: **The Majesty and Power of The Pure Truth of His stupendous and humanly-incomprehensible Creation**.

Fenton has therewith bequeathed to the worlds of science and religion the ordained foundation upon which to build, within their Disciplines:

The 'Harmonising Truth' about Creation and Evolution;
and thus the True Origins of Man!

His translation and insights offer far greater enlightenment than do other mainstream Bibles around the ongoing, contentious and quite foolish debate surrounding the *Creation/Evolution* paradigm. Therefore: Among the very many Bible translations *inundating* global Christendom, **Ferrar Fenton's Bible** stands as *the strongest overall*.

Preface

The subject matter of this Third Edition Booklet is now
substantially derived from Chapter 2 of the *primary* Parent
Work:
BIBLE "MYSTERIES" EXPLAINED:
Understanding "Global Societal Collapse" From The
"Science" In The Bible;
What Every Scientist, Bible Scholar and Ordinary
Man Needs to Know.

Given the sometimes acrimonious history of the Creation versus
Evolution debate, the clear and strong sub-title stating that "Genesis and Science" agree will, I am sure, cause many on either side
of the debate to be derisively dismissive of such a claim. Nevertheless, the very fact that we human beings are actually living and
breathing on the earth unequivocally illustrates that irrespective of
whichever view one might personally choose, the division itself is
ultimately rendered irrelevant, *precisely because* of our very presence here. For we arrived, we are here.

So what is the debate really all about? Could it ultimately be
about protecting ideas so long considered sacrosanct that pride and
perhaps ego will no longer even consider any kind of accommodation whatsoever?

Obviously we should seek to know our origins, for such knowledge permits us not only the *"where from"*, but also the *why*
and the *how!* While all three aspects naturally offer the complete
picture, the key 'why' singularly allows us to focus on what we as
a collective humanity have produced on our blue planet home over
the many thousands of years of our tenure here. The picture is not
a good one.

[**Note:** The crucial reason for the *complete* and absolute *why* will
be found in the Parent Work.]

It would be fair to say that for the most part science, particularly "Anthropological Science", would shy away from any notion that Creation, as it is currently accepted and promoted by the mainstream Christian Churches, could find accommodation with their views. Therefore, the solidly entrenched beliefs of Evolutionists and Creationists appear to be so diametrically opposed that attempting to show the opposite from The Book of Genesis would, on the surface at least, seem futile. Yet that is exactly what we can reveal, and therefore what we will do.

The key lies in reading Genesis very, very carefully line by line, with the fundamental and crucial requirement to really understand the key difference between what is explained in Chapter 1, and what is revealed in Chapter 2! A critical examination of the meaning of, and connection to, every other line is therefore imperative for complete understanding. Obviously, all assumptions and preconceptions should be discarded before undertaking such an analysis. One should also examine a number of different Bibles at the same time for there are translations which, in their interpretations, mask the great revelation and knowledge about the science behind Creation in Genesis.

Of course we do not, in any shape or form, place human science before the great Creative Process. In truth, the nature of the process as described in The Book of Genesis cannot ever be fully understood by either science or religion – simply because it describes an event utterly stupendous and incomprehensible, originating out of an Eternal, immeasurably-high non-material sphere or level, to its lowest precipitation in the material worlds of the physical universes. Incomprehensible as it is, however, it is nevertheless still the role of earth sciences to at least strive to understand that singular key **'Event'**! For it brought forth the worlds necessary for life material – which, paradoxically, provides us with the very means to *discuss* that exact happening.

The time-frame that the concept of Creationism espouses for human entry onto the earth is also clearly at odds with Evolutionism and probably represents the greatest sticking point in this whole debate. The 4.5 billion years that Cosmology accepts versus the 6000 years of fundamentalist Christianity is clearly insurmountable. As I have stated a number of times in the Parent Work, *either a thing is so, or it is not.* It cannot be both at the same time. It certainly cannot in terms of the huge unbridgeable difference here.

Notwithstanding such an obvious reality, is it possible for one key aspect of the debate to be immediately and perfectly reconciled? In other words, can Evolution actually be Creation? Can the Creation Process, in its actual outworking to produce the great universes, at the same time also be the Evolutionary process? If so, then the apparently irreconcilable and huge time difference is automatically resolved too.

One noted scientist has publicly stated what we unequivocally affirm; that Evolution and Creation are mutually inclusive; that they are virtually two sides of the same coin.

Francis S. Collins, M.D., PhD, director of the National Human Genome Research Institute and author of the book, "The Language of God: A Scientist Presents Evidence for Belief", in a recent interview on "Anderson Cooper 360°" (CNN), had this to say:

> "As the director of the Human Genome Project, I have led a consortium of scientists to read out the 3.1 billion letters of the human genome, our own DNA instruction book. As a believer, I see DNA, the information molecule of all living things, as God's language, and the elegance and complexity of our own bodies and the rest of nature as a reflection of God's plan, I did not always embrace these perspectives. As a graduate student in physical chemistry in the 1970's, I was an atheist, finding no reason to postulate the existence of any truths outside of mathematics, physics and chemistry. But then I went to medical school and encountered life and death issues at the bedside of my patients. Challenged by one of those patients, who asked: "What do you believe, doctor?", I began searching for answers. I had to admit that the science I loved so much was powerless to answer questions such as: "What is the meaning of life?" "Why am I here?" "Why does mathematics work, anyway?" "If the universe had a beginning, who created it?" "Why are the physical constants in the universe so finely tuned to allow the possibility of complex life forms?" "Why do humans have a moral sense?" "What happens after we die?"

Dr Collins was also asked questions by others, such as:

> "Can you both pursue an understanding of how life works using the tools of molecular biology, and worship a creator God?" "Aren't evolution and faith in God incompatible?"

Dr Collins continues:

9

> "Actually, I find no conflict here ... evolution by descent from a common ancestor is clearly true. If there was any lingering doubt about the evidence from the fossil record, the study of DNA provides the strongest possible proof of our relatedness to all other living things."

Proceeding from his remarkable statements, Dr Collins's following insights are precisely those which are inherently imbued with precise Truth, and which unequivocally concur with the explanations offered in this Booklet.

> "But why couldn't this be God's plan for creation? I have found there is a wonderful harmony in the complementary truths of science and faith. **The God of the Bible is also the God of the genome.**"

> (Emphasis mine.)

The following 'essay' thus offers a very different look at a subject that, for many hundreds of millions, is simply intractable in its "seemingly" unbridgeable chasm.

Whither Cometh Humankind?

> "When we consider thy heavens, the work of thy fingers,
> The moon and the stars, which thou hast ordained,
> What is man, that thou art mindful of him?"

(The Gospel of the Essenes.
E. B. Szekely, p.175.)

"Where does the world come from?"

"She hadn't the faintest idea. Sophie knew that the world was only a small planet in space. But where did space come from?

"It was possible that space had always existed, in which case she would not also need to figure out where it came from. But could anything have always existed? Something deep down inside her protested at the idea. Surely everything that exists must have had a beginning? So space must sometime have been created out of something else.

"But if space had come from something else, then that something else must have come from something. Sophie felt she was only deferring the problem. At some point, something must have come from nothing. But was that possible? Wasn't that just as impossible as the idea that the world had always existed?"

(Sophie's World, Jostein Gaarder, p.8,
Phoenix Press.)

The questions that young Sophie finds herself faced with in Gaarder's International best-seller are the very same that men have asked since time immemorial. The fact that this particular publication was a "best-seller" illustrates the keen, ongoing interest to such ideas from far more than the scientific or philosophic community. For, quite logically, there should be a natural, inherent curiosity in each of us which brings forth these very same questions by virtue of the fact that we exist on planet earth in the first place. Let us join Sophie as she asks more questions!

> "How was the world created? Is there any will or meaning behind what happens? Is there a life after death? And most importantly, how ought we to live? People have been asking these questions throughout the ages. *We know of no culture which has not concerned itself with what man is and where the world came from.* But history presents us with many different answers to each question."

> (Sophie's World. p.12, Italics mine.)

Sophie's second question is addressed in Chapter 3 of the Parent Work from the standpoint of The Spiritual Laws.[1] This, in turn, provides the answers to the fourth question, namely how we should live. The third question on 'life after death' is also addressed in the Parent book. For if we do not know from where we originate, we cannot know who we are, what we are, or what our purpose is. That leaves only the first question to be answered, "How was the world created?" That question we examine here – **in this Booklet!**

The philosopher Plato believed that there had to be a reality behind what he termed the "world of ideas". He thought that we could never have true knowledge of anything that is in a constant state of change. We can only have true knowledge of things that we can understand with our reason. Plato also believed that man is a dual creature with a body bound to the world of the senses – which he thought were unreliable – and an immortal soul which is the realm of reason. He also believed that the soul existed before it inhabited the body.

Even though Plato wrote extensively about this dual concept, it was widely believed by many Greeks before him. Plotinus (ca. 205-270 AD) knew of similar ideas from Asia. Plotinus also believed that the world is a span between two poles with The Divine

---

[1] Also addressed in a 'Sister Booklet': **"The Spiritual Laws of Creation** – *The Crucial Knowledge for Humankind"*.

Light at one end, and absolute darkness at the other which receives none of The Light. He believed that this darkness was simply the absence of light, without any existence. Thus The Divine Light becomes increasingly dimmer the further one travels from it. Finally, he believed, there is a point that it cannot reach. He believed, moreover, that the soul is both a spark from, and illuminated by, The Divine Light; a fascinating insight because this particular view contains much basic truth, as we will illustrate.

Whilst the philosophers have mulled over this question for millennia, the mainstream religions have generally not needed to do so. The acceptance of an immortal soul or spirit – whether it becomes "one with the universe" or retains its "personal self-conscious form" after earthly death – is part and parcel of most religious beliefs. Even within this field, however, there is no clear position either. Yet pre-dating the early philosophers and the main religions, we find that in some places "cave-men" buried their dead with flowers and small items that were probably personal possessions.

This indicates that the funeral ceremony was conducted with a certain ritualistic air, perhaps reflecting the first stirrings of a belief in the duality of man in the early progenitors of humankind. A degree of reverence for either the burial process or in the belief of a soul departing from the body is evident here. This offers a different perception of these early humans, once stereotypically depicted as brutish.

Democritus (ca. 450-370 BC), on the other hand, believed that people and animals were constructed solely of atoms, and that neither possessed immortal "souls". According to him, souls were built up of atoms that are dispersed to the winds when people die. In contrast to that particular view, **The Spiritual Laws** *of* **Creation**, as the driving power throughout *all* **Creation**, decree that there are both animate and inanimate life-forms. The inanimate we may designate as that which is anchored in place such as trees and mountains. Rivers, lakes and glaciers etc., also come into this category. The animate is naturally the opposite and comprises those life-forms that are, in essence, *mobile!* They include the insects, birds, fishes, animals and, of course, man.

The "mobile group" is further divided into those forms that have free will and those that do not. In fact, only man possesses free will. The *mobile life-forms* of the natural world do not. They do, however, possess instinct. The designation "mobile" means that all such creatures possess an "inner animating core" separate from their "physical form". We can picture that "inner core" as

13

a "power pack"; the "battery in the machine", so to speak. However, because there is a huge and fundamental *inherent* difference between the *free will* of humans and the basic *instinct* of all other mobile life-forms, there is, similarly, the same great difference in the *respective kinds of animating power* contained within the "individual species".

For a greater understanding of this essay, the concept of free will essentially requires deep thought. Because its nature has puzzled thinkers for centuries – free will being inherently necessary for any decision-making process – the non-understanding of what free will actually is and how it works effectively *prevents* the understanding of the many problems that beset us. Problems appear which *seem* to have no causal reason at times.

Immanuel Kant, (1724-1804) a German idealist philosopher, Protestant and an ethical man, believed strongly in three things; that man has an *immortal soul*, that God *exists*, and that man has a **free will**. These aspects, he believed, were essential factors under which the necessary virtue of morality could flourish. By exercising free will in his decision-making process, man moulds and shapes his individual personality in accordance with the strict outworking of certain immutable Laws, thereby determining his future. The subsequent "level of development" attained is then his alone. This simple yet absolute mechanism explains how inequalities occur – why men are not equal.

In order to fully understand our true nature as human beings, it is important to also understand what exactly constitutes the three main parts of the complete entity, earthman. In simple terms, they are the material or *physical* body, the *soul*, and the *spirit*. [The *mind and emotions* are part of the *physical* aspect because their contribution largely stems from the activity of the brain.] Unfortunately, the designations *"soul"* and *"spirit"* either cause great confusion in the differences *not* being understood, or are thought to be the *same* thing.

Essentially, we can designate "spirit" as being the **innermost animating core** of man. His "spirit" is that "primary aspect" which inhabits the material body. ***It is the actual animating power or force.***

The "Spirit" is thus the *actual* person!

The soul body and physical shell are respective ***outer coverings*** that ***clothe*** the "spirit". We may thus regard the soul as being composed of all the *"other-world" coverings* that *envelop* the **spirit**. Obviously, it is *not that* which is the **material** body. [In the Ethereal World of the *beyond*, however, it is the *ethereal body* that envelops the spirit. For the material body at this time has been vacated.]

To reiterate; The Spiritual Laws therefore decree that human beings have both a physical body and a non-material, "other-world" one collectively called the *soul*. The soul, in turn, envelops the inner essence of who and what we actually are – our **spirit**. Therefore, *we are spiritual beings first and foremost*. So, whilst *spirit and soul* are closely linked together inside man's physical form as *part* of the complete entity, by virtue of their *different origins* they serve a slightly *different purpose*, even though still being *animating aspects* of a *singular* human entity. Therefore, in addition to knowing the process by which we arrived here, what we also need to know is the how and why of our spirit, the how and why of our various "coverings" of the "soul-body", and finally how they all fit together.

Now, whilst man possesses ***Spirit*** as his "innermost animating power", this is not the case with animals. Their "inner life force" may be designated as being *soul only*, because their ultimate place of origin stands at a *lower level* in Creation than man's *higher level* of Spiritual Origin.

Animals, therefore, do not possess the *spiritual* responsibility *inherent* with a free-will attribute as does humankind, and are thus not subject to the reciprocal outworking of particular, immutable Laws as humans are. Unfortunately for them, however, they *are* subject to every whim of mankind.

From those brief explanations it can be safely deduced that *inanimate objects* cannot therefore possess *either* soul or spirit. What is sometimes *perceived* or *felt* around great trees and around waterfalls or in mountains and forests is something entirely different, which in a sense can be called "soul" but is not of the same kind.

So, having determined the nature of free will, soul and spirit – the origins of which we will discover in this Booklet – the next vital step is to address the question of the Creation-process itself. Since there has never yet been complete agreement about the "origin of the world" and all that it contains, let us add our voice to the debate and offer our explanations for mankind's beginnings to try to resolve the first question that Sophie is struggling to come to terms with. The reader will thus discover for himself that the following explanations do finally provide clear and *logical* enlightenment to this *seemingly* perplexing question.

Firstly, in concert with the discoveries of anthropological science, traditional, cultural and indigenous beliefs which state that ancestors were "things" or "beings" other than human, should be dismissed. Human beings are not descended from rivers or mountains or half-men or demi-gods that some indigenous beliefs allude to. Nor from any other form that superstitious conviction or legend has dreamed up, and neither are any races different with respect to the truth of this. Interestingly, this is not a religious truth, or even a "scientific" one. It is the Truth simply because the human form is the **only one** ordained for all of humankind under the outworking of **Spiritual Law** – that Law which overrides and transcends all man-made beliefs.

Let us, therefore, begin with an emphatic and bold statement! Those souls presently living on planet earth have, as their common Spiritual Origin and therefore their true home, *The Spiritual Realm of Creation!*

What do we mean by this, and what is The *Spiritual* Realm of Creation? Is there such a place? Can there be such a place? If there is, are we possibly "related" to all other races on earth by virtue of our same place of Origin? If so, were we once all there together as **different races** as is the case on earth, or are those differences "recent developments" which permit reference points by which we can trace natural, physical and regional differences? And do we all automatically return to our place of origin at earthly death a member of "our own particular earthly race" as a matter of course, or are there other possibilities?

Where and how do we find the answer to this most necessary of questions – our Origins? For there is surely little point in journeying through life uncertain, confused, even angry at vexing questions such as race, racial mix and ethnic origins or whether one truly belongs to one race or country more than to any other. So, in the context of this particular composition, where should we begin?

We know that most races, cultures and religions have, as a common theme, a story of "Creation". The myriad views expressed in cosmology and Creation-theory etc., are as diverse and as numerous as the thought-processes that have spawned them. Yet very few of the many thousands of varying ideas completely agree with each other, and will therefore not be completely correct in their *entirety*. Certainly many aspects may be similar and have elements of the truth contained within them. It would be equally true to also say that in the natural process of evolution and development, certain races would have garnered insights which would have required previously accepted beliefs of "truth" to be discarded, in the sure knowledge that the "new" contained more "truth" than the "old".

The transition to the "new" was not necessarily without great travail and societal and religious upheaval, for man's precious ego does not easily allow him to "let go" what he considers to be his "great well of truth". Yet in the major issues relating to our origin; to correct living according to Spiritual Law and to our entry and exit from earth, the greater mass of humankind today remain just as ignorant and without real knowledge as they always have.

Notwithstanding the inherent correctness of that statement: Through the Wisdom contained within The Creative Will, certain especially chosen ones *still unveiled The Truth to humankind* at precise points in history in accordance with the developing spiritual maturity of the particular people into whose midst they were incarnated. Those teachings were given to mankind in a manner that could be understood by the race or nation that had been specifically prepared to receive it from the "Truth-bringer" incarnated for *that particular people* at the appropriate time in *their* "spiritual" journey. This took place over many thousands of years. Man, however, could not leave the Teachings alone. He had to alter the original clarity of them.

In essence, the reluctance to discard precious pet beliefs reflects the inexplicable inability of mankind to leave such teachings in their pure and uncorrupted state. Rather than simply *living* the teachings as instructed, man chose to dissect them to suit his personal wants. This unfortunate fate has befallen all the great spiritual teachings – **without exception**. Even the setting up of the major religious institutions of so-called "higher-learning" has not advanced the cause of *Truth* a great deal.

In reality, its unfortunate "dissection" has been directly responsible for the incredible proliferation of so many *different religions*, with each generally purporting to be the only true one. After being offered the great Truths, man subsequently converted them all to *"just religions"*. In a kind of perverse irony, if all the Teachings had been lived in a purely unadulterated way by those peoples to whom it was given, there would now only be *one single unified Teaching throughout the world today.*

And that – as stated by **The Son of God** – would have been:

The "All-Truth!"

The German philosopher, Hegel (1770-1831), even though admitting to the existence of an "unattainable truth", believed that "truth is subjective" and "all knowledge is human knowledge". He thus rejected the existence of any "truth above or beyond human reason". He believed, moreover, that because human ideas changed from one generation to the next, there could not be such a thing as "eternal truths" or "timeless reason". In his view history provided the only fixed point that philosophy could cling to. Since history in the philosophic sense is more or less constant reflection, Hegel believed that certain rules applied for this "chain of reflections".

Thus a thought is usually proposed on the basis of other, previously proposed thoughts. However, the proposal of one thought is invariably contradicted by another, thereby producing tension between the two opposing views. That is basically the current position with regard to the Creation versus Evolution debate. Hegel postulates a method whereby such entrenched positions can be "softened" so that the tension is resolved by the proposal of a third thought which accommodates the best of both points of view. Hegel calls this a *"dialectic process"*.

Unfortunately, the belief still persists that science and religion will probably never be truly reconciled, and that perhaps they should not be. However, since our purpose is to move from a simple faith/belief position to genuine conviction as to the truth of it all, the knowledge contained within The Spiritual Laws perfectly *permits* a reconciling of the two viewpoints. We will, therefore, in this crucial "journey", utilise Hegel's *"dialectic process"* to identify which of the respective *main points* of each argument can offer *mutual accommodation* without losing any genuine substance from *either*, thus merging the *two* into *one complete and logical whole*!

Classically, the subject of Creation has provided the perfect forum for completely opposing views – that of orthodox Christianity and Creationism against that of the scientific community generally tending toward Darwinism and/or Evolution. Galileo, Darwin and others whose findings challenged Church dogma were invariably branded heretics, and the polite way to reconcile science and theology was simply to agree that each would keep to one's own area. Basically, science would ask and answer the questions what, why and how empirically; and the church would do the same from the religious or spiritual standpoint.

In April 1997, what was billed as the Great Noah's Ark Trial was held in a Sydney court. Whilst not a case of an "Evolutionist" versus a "Creationist" in the classic sense, Dr Peter Pockley, a freelance journalist, nevertheless reported that "...the trial had pitted the belief of many fundamentalist Christians in the literal truth of the poetry in Genesis against the conclusion of science for a 4.5 billion year old earth."

In reality, there is no conflict between the two positions, as we will illustrate. The disagreement exists only in the minds, and therefore in the *incorrect interpretations*, of the proponents of the respective points of view. Moreover, essentially *the same battle*, which we note further on, was fought in an American State Supreme Court years ago.

So even with the current strong corporate-earth mindset today, the age-old question of "man or monkey" still produces passionate debate. Since this essay offers clarification of that debate, and with what we have at our disposal, let us now employ Hegel's *"dialectic process"* to determine where such truth lies in discovering an origin for ourselves that makes sense. The positions of science and religion should thus blend harmoniously, each supporting the other without conflict as it obviously should be. For we human beings of earth are the proof of this one simple reality.

We exist! We are here!

In the evolutionist's corner, Charles Darwin (1809-1882) – once described as the most dangerous man in England because of the direct challenge to the teachings of Christian orthodoxy that his work of evolution brought – proposed that "...all existing vegetable and animal forms were descended from earlier, more primitive forms..." via the simple mechanism of biological evolution. And that evolution was the result of "natural selection". Until quite recently science had "pushed back" and accepted a geological "birth-date" for our Solar System and the earth as approximately 4.6 billion years. Darwin thought the age of the earth to be about 300 million years. The age of the universe itself is believed to be around 14 billion years or so.

Historical anecdotes of Darwin's ideas possibly being correct actually sent shock waves through the "establishment", with even a distinguished scientist noting that it was "an embarrassing discovery", and "the less said about it the better". An "upper-class lady" expressed the hope that it was "not true", but if it was, then the further hope that it would "not be generally known".

> *"Much of the vitriol directed at Charles Darwin a century and a half ago came not from his ideas about evolution in general but from his insulting but logical implication that humans and the African apes are descended from a common ancestor. ... Along the way they [palaeontologists] learned, among other things, that Darwin, even with next to no actual data, was close to being right in his intuition that apes and humans are descended from a single common ancestor – and, surprisingly, that the ability to walk upright emerged millions of years before the evolution of our big brains."*

(Time Magazine, Oct. 9th, 2006.
"How We Became Human".)

So, standing in the opposite corner to Darwin stood the Creationists and Genesis "literalists". In Darwin's time both the ecclesiastic and scientific views were virtually sacrosanct with regard to the doctrinal idea that all vegetable and animal species were created only once in each and every respective form. The views of Aristotle and Plato were not dissimilar to the Christian beliefs, since they basically thought that all animal species were patterned after "eternal ideas". This Creationist outlook, in concert with the Biblical, genealogical time-frame back to "Adam", postulated that the earth was "created" about 6,000 years ago.[2]

In determining the various arguments for the "Creation versus Evolution" debate, and in the context of the subject matter in this work, it is vitally important to know what evolution means – exactly. There is a view in some scientific circles that evolution means "selection by random chance", and not perhaps to a precise developmental path. British Astronomer, Sir Fred Hoyle, stated that "...believing that the first cell originated by chance is like believing a tornado could sweep through a junkyard filled with airplane parts and form a Boeing 747." Professor N. Chandra Wickramasinghe, co-author with Sir Fred Hoyle of "Lifecloud: the origin of life in the universe", has also dismissed the evolution idea.

Overall, however, the science of astronomy believes it can trace the evolution of the universe – "...on the assumption that matter is created; but just how it is created is another problem altogether, and no theory has given indication of how this came about." (The

---

[2]In fact, using so-called biblical genealogy James Ussher, a 17th century Bishop, calculated that the earth "began" at 6pm, October 23, 4004 BC.

Atlas of the Universe, "Origin – or Evolution", p.214) Yet some scientist/theologians believe that evolution provides clues to the very nature of God.

Plato, (428-347BC) who was basically concerned with what was eternal and immutable on the one hand and on what "flowed" on the other, found mathematics very absorbing because "...mathematical states never change." Much later in the seventeenth century Galileo observed that the book of nature "...was written in the language of mathematics." "Measure what can be measured, and make measurable what cannot be measured."; was his view.

The actuality of Immutable Laws, and therefore a Creator of those Laws which *automatically govern* "His Creation", negates the idea, for example, that life on earth could ever have been the result of "random-chance" development. This image or notion does not "hold up" because the Creative process, under the outworking of The Spiritual Laws, translates that very mechanism into *precise mathematical formulae* in the material spheres. Thus **The Spiritual Law of Numbers** is also mathematical Law which can be noted in everything, everywhere.

Even the "primordial soup", produced at the birth of our planet aeons ago, had to have the appropriate formulae out of which eventually developed all the physical life forms and substances for planet earth. Within each will be found their own *personal-species mathematical formula.* Change the formula and you change the substance or thing; if such change can be achieved within the bounds of scientific law which, in reality, is Spiritual Law.

Since this Law provides for development, – but not for alteration outside of what is possible – science, therefore, *cannot actually create anything new at all.* It can, however, produce *new combinations*, but only from substances that *already exist.* Even then, only *within* the parameters of what is scientifically and therefore spiritually possible under The Creative Will. It can, of course, discover things not previously known to *earthly humanity.*

The Bible alludes to this mathematical precision in the exactness of The Laws of Creation by stating that everything is counted and nothing goes unnoticed. As a simple illustration of this lawful truth, scientific formulae decree that only a precise number and configuration of certain atoms can form molecules of a particular substance. Change the number and a different substance is produced, or the experiment may not work.

For example, one atom of copper, one of sulphur and four of oxygen will combine to produce $CuSO_4$, which will forever be copper sulphate. Copper sulphate, quite logically therefore, cannot ever be $CuSO_6$ or $CuSO_9$. In the same way common salt, chiefly sodium chloride, will always be $NaCl$, not Na_2C_{16} or 8 or any another formula. And sodium bicarbonate, commonly used in the kitchen in the form of baking soda, can only ever be $NaHCO_3$, not Na_5HCO_7 or anything else.

Physicists have noted signs that the cosmos is custom-made for life and consciousness. It turns out that if the constants of nature – unchanging numbers like the strength of gravity, the charge of an electron and the mass of a proton – were the tiniest bit different, then atoms would not hold together, stars would not burn and life would never have made an appearance. John Polkinghorne, a former distinguished physicist at Cambridge University and now an Anglican priest, sagely observes: "When you realise that The Laws of nature must be incredibly finely tuned to produce the universe we see, that conspires to plant the idea that the universe did not just happen, but that there must be a purpose behind it."

Charles Townes, who shared the 1964 Nobel Prize in physics for discovering the principles of the laser, goes further: "Many have a feeling that somehow intelligence must have been involved in The Laws of the universe." And the authors of "The Mystery of Life's Origin", concluded that: a "...**Creator beyond the cosmos...**" *is the most plausible explanation of life's origins.*

Colin Patterson, senior paleontologist at the British Museum – once accepting of notions similar to Fred Hoyle and N. Chandra Wickramasinghe – after believing in evolution for more than 20 years, claimed he was "duped". Charles Darwin seemingly noted that not one change of species into another is on record, and he could not prove that a single species had been changed. In 1984, the former President of the French Biological Society, Professor Louis Bounovre, stated: "Evolutionism is a fairy tale for grown ups."

On the other hand, Arthur Peacocke, a biochemist who became a priest in the Church of England in 1971, has no quarrel with evolution for he finds in it signs of God's nature. He infers from evolution that God has chosen to limit His Omnipotence and Omniscience. In his apparent view, it is the appearance of chance mutations and the Darwinian laws of natural selection acting on this "variation" that bring about the diversity of life on Earth. Theologian, John Haught, founder of the Georgetown (University) Centre for the Study of Science and Religion, believes this process suggests a Divine humility, a God who acts selflessly for the good of Creation.

The sticking point in this whole debate is perhaps not actually that of Creation versus Evolution in any case, but probably more that of the time-frame required to produce either one, or both together! For if a time period for such a thing as "evolutionary-Creation" can be logically established then both viewpoints can be accommodated in perfect harmony. In my view, therefore, the one key question in this debate that must be considered – *yet rarely is* – is: **"Can Creation also be evolution?"** And/or vice-versa? Our reply is an unequivocal:

Yes, it can! And, moreover, it is!

The simple recognition that "Evolution" can actually be a necessary and vital part of the "Creative-process" should logically develop into the clear conviction that Evolution *is* the Creative-process. At the same time, the Creative-process *is* Evolution. That is fact, simply because The Will of God must inherently be both natural and logical in all processes. That fact also inherently stems from the Perfection of His Creative Will. Therefore, since it is vitally important to understand what we actually mean by "evolution" – or what it is supposed to mean – from our particular standpoint we shall state it to mean, and also encompass and promote, "natural development"! We do not mean "random selection", or anything even remotely equating to any kind of "chaos theory" either.

Unfortunately, the respective default settings of *science and religion* presently seem to be so irreconcilable that there would appear to be no grounds anywhere for a meaningful merger of both realities. On the one hand science seems generally to hold to a kind of *eternal doubt* paradigm whilst the basic core of religion is *faith*. Since a major sticking point remains the possible time period necessary to accept both the scientific and religious or theological viewpoints, these two contentious aspects will nonetheless be drawn together to show that such a merger is not only possible, *but is actually the true position* – notwithstanding the many interpretations to the contrary from proponents of both disciplines.

To begin with, the very first thing is to revisit and clarify exactly who and what we are. If we unequivocally state that our true origin is that of The Spiritual Plane of Creation, we logically imply that we cannot be purely physical in origin, just as the early Greek philosophers surmised. Therefore it may be presumed that the earth plane of the material world is *not our true home* but a *material one* for the time that we are ordained to live on it. Yet the apparently obvious reality appears to indicate the opposite. We see, hear and feel everything around us as being solid and material. To all intents and purposes it seems logical to believe that we are **solely** material beings. Should we expect that to be the last word if our "Spiritual Origin" is **not that** of the world of matter, however? The key to it all lies in understanding this *dualistic reality*!

If we examine The Law of Gravity we observe that heavier objects sink whilst lighter substances rise. Whilst being obviously so, it is nevertheless important to stress this point before proceeding with our explanations. Thus, in the structure of Creation, a *material* object will occupy a lower level or plane than a *spiritual* one, simply because the higher the Plane of Creation, the finer and *lighter* is the substance of which it is composed. The same principle applies for the particular inhabitants of those respective Planes.

Therefore, only the earth of the material plane is the home of the flesh. In this material world we marvel in awe at the vast expanse that we see in the night sky. Astronomers speak of interstellar distances so incredibly immense that the human brain can scarcely even begin to comprehend such figures. We are reminded of the magnificence of such a work through the insights of the poets and philosophers in their attempts to understand our place in the universe:

> "When we consider thy heavens ... what is man, that thou
> art mindful of him?"

A powerful question, indeed, and one we should all ask of ourselves from time to time. For in the stupendous scale of things, it is vital to understand that the mind-numbing, incomprehensible immensity of just the physical universe alone can only be the **smallest and lowest part** of the **whole of Creation**. Our galaxy, on its own, contains something in the order of 100,000 million stars. The known universe, in turn, contains literally billions of such vast galaxies. The higher Spiritual Planes, by virtue of their far greater spiritually-expansive attributes, are therefore incomprehensibly and immeasurably *far more immense*.

If, then, our Origin is out of The Spiritual, yet our earthly existence is obviously material, the only conclusion that we can logically draw here is that we must have both these aspects contained within us, i.e., the human being is both spiritual and physical.

And that is so! Only on the earth, however, can this duality be utilised. Indeed, it is the only way that humankind can meaningfully exist here at all.

> ***For the spirit needs the material body to fulfil its purpose whilst on the earth, and the body needs the power of the spirit to animate it here also – to give it life.***

Earthly death, therefore, is little more than the separating out, the drawing apart of the two; where also at this time the "spirit" *should* strive to free itself from the material world. The material shell then returns to the earthly components from whence it came in the normal process of decay that The Laws of Nature decree must take place. Hence the words of The Law: "Earth to earth, dust to dust..." which we hear at funerals, and which only applies to the empty, discarded shell. The process is simply outlined in Job 10:9-12:

> "Remember You made me from clay,
> That to dust You will make me return!
> And did You not curdle the milk,
> And fixed me together like cheese,
> Then clothed me with skin, and with flesh,
> And with bones and with muscles compact?
> And gave me my life and my reason,
> Then last, *fixed my Spirit in me*?"

> (The Holy Bible in Modern English.
> Ferrar Fenton. Italics mine.)

Accompanying the process of establishing the physical/spiritual connection is the requirement to locate relevant reference points pertaining to the respective origins of both those parts to man in order to gain the necessary understanding. We will thus thereby learn why, in the development of man and transition to human, (us) the spirit was *fixed last* – as Job states.[3]

Therefore, of all the *religious works* that *purport* to have the Truth – insofar as most Western peoples are concerned anyway – The Bible is probably the best known and accepted by virtue of

---

[3] Those key points are derived from certain, crucial, now *mainstream* writings.

25

the fact that it is the one where the person of Jesus Christ is the key figure. Because of His particular Origin and pivotal role for humankind as documented in The Bible, we will therefore accept this book – at least among the *religious* works – as being *more able to provide* the answers we seek.

Thus, in one single, simple sentence from The Bible, both our *physical* and *Spiritual* Origins are actually *clearly* revealed. In its stupendously far-reaching yet stunning simplicity, it **completely destroys** the "great divide" that religion and science have constantly promoted and clung to. We ask why?
The King James Version of Genesis Chapter 2 Verse 7 states:

> **"And the Lord God formed man of the dust of the ground, and breathed into his nostrils the breath of life; and man became a living soul."**

(Emphasis mine.)

Here has lain **one part of the answer**, for centuries unnoticed, unseen perhaps, but clearly not at all understood.

Now, therefore, if this particular Scripture is thought about in purely literal terms or from solely a fundamentalist viewpoint, a picture more or less naturally arises of The Creator, The Power of All that exists, *descending* to earth and building the shape of a man out of its substance – the mud of it. Then, what would effectively be a *model* of a *mud-man* would be instantly transformed into a living, breathing, walking, talking, internally-pulsating human being by the simple act of being *"breathed into"* in the *literal* sense.

Is that the method by which one could believe man was first formed? The crudeness of such an idea is difficult to reconcile with a Creative-force responsible for the Creation of not just the incomprehensible immensity of the physical universes alone, but the far larger and far higher spheres spoken of in many religious works and attested to by **The Son of God Himself**. With all that we have learned about our multi-faceted world today, is there any point in continuing to cling to such a preposterous idea?

Moreover, there is a major and insurmountable problem for literal fundamentalist thinking here in that The Bible alludes to the fact that God **cannot** *descend* to the earth for it would be completely consumed by His Power. Immediately there is a contradiction, if we view it in a purely literal sense. Quite unequivocally, therefore, there can be **no** contradictions anywhere in the *actual Creation process itself.*

We should thus be very careful not to apply any kind of heretical or blasphemous labels to an idea that may be markedly different to any current or orthodox Church one. We should, instead, objectively allow the intuitive inner reason the spiritual freedom to determine the true nature of what is revealed here. Perhaps a wider vista might then suddenly open up offering the spirit the potential to "soar" instead of being shackled by too rigid an interpretation that refuses to allow even the "possibility" of such a thing as "Evolutionary-Creation".

The much-celebrated "Tennessee monkey trial" or "Scopes monkey case" of 1925 in Dayton, Tennessee, provided the key forum for exactly this debate. A high school biology teacher by the name of John T. Scopes who taught the theory of evolution, was accused of violating the Butler Act, a Tennessee law that forbade the teaching of Evolution because it contradicted the account of Creation in The Bible.[4] The trial received worldwide publicity and was conducted in a circus-like atmosphere. Because of the popular belief that evolution meant humans were descended from monkeys, the press dubbed it the "Monkey Trial".[5]

Because of its far-reaching educational implications, not least for many of the scientific disciplines, the Education Department hired the famous criminal lawyer Clarence Darrow as their Defence Counsel, whilst a former US Secretary of State, William Jennings Bryan, appeared for the prosecution. Clarence Darrow and his team argued for the scientific validity of evolution and against the constitutionality of the Butler Act. According to anecdotal reports, and after both views were aired, the case hinged on one crucial question which Darrow addressed to the opposition. The question concerned the existence of dinosaurs and the time-frame in which they lived.

Since their existence could not be denied, the challenge and case could not be upheld. Had fundamentalism won that day, the State Supreme Court would have had no option but to order schools to teach only "the 6 days-of-Creation belief". Any concept of evolution would have been officially suppressed. The Butler Act remained on the State Statute books until 1967. Paradoxically, that court case need never have taken place simply because, as is our premise, the evolutionary process is actually part of the Creative

---

[4]Trials along the same theme have taken place in the USA even in more recent times.

[5]The similarities between man and the anthropoid apes evidently caused Darwin to believe that both probably evolved from the same progenitor.

process, and naturally so.

*Indeed it could not possibly be anything other than a **natural union**.*

Yet, even in the 21st century now, the science of "biological anthropology" still persists with the totally incorrect belief that it is *solely genetics* which has determined the so-called "evolution" of "primate to human". Researchers from the Broad Institute of MIT and Harvard have evidently coined a new "anthropological term" – *the human-chimp split*. Their *basic* hypothesis and time-frame *is* correct. An ancestral ape-species *was* the ancestor of the human race a very long time ago, **"but only as the physical-form vessel, nothing more!"**

Science must recognise and learn to understand the huge and fundamental difference between that *physical/material-form of vessel/body*, and *the animating life-force within it*; i.e., within *every human being*. Only with that essential knowledge as the primary foundation for any further research – if it is deemed necessary – might the current, strong scientific emphasis on genetics and the human genome find its *correct connection*. Rather than constantly needing to change hypotheses, this particular branch of science might, instead, begin to build *constant* upon *constant*.

So if we revisit our previous Genesis quote from the King James Bible, in Chapter 2 Verse 7 we discover therein the amazing revelation of *the actual human/chimp split*.
That key Verse states:

> **"And the Lord God formed man of the dust of the ground, and breathed into his nostrils the breath of life; and man became a living soul."**

(Emphasis mine.)

Therein lies the *overarching answer* to the most amazing evolutionary processes that, in the most natural and logical way, *separated out the first human beings from their physical-form progenitors.*

Thus, The Bible described **the real human/chimp split** a very long time ago.

Most unfortunately for the *correct* education of the many millions of students globally who are taught otherwise; in a fascinating paradox, *both* human *science* and human *religion* have not yet caught up with this fundamental and key, inviolable Truth. In the *current* educational paradigm, moreover, they are not ever likely to. The necessary change, however, *will occur* as more and more *previously-sacrosanct* religious and scientific tenets increasingly come under the very **Spiritual Pressure** now forcing once proud notions, beliefs and institutions to collapse.

The detail of this singularly-decisive event for humans more fully unfolds as we continue our journey, and clarifying explanations flesh out the primary aspects.

Despite the incredible nature of what has been revealed here, it is still only a part of the complete process, albeit a stupendous one for human beings. A far greater revelation has been present in The Bible since its inception.

0.1   The "First" Creation

That *greater* revelation is written in The Book of Genesis, most relevantly of course in Chapter 1. Verse 26 therein states:

> **"And God said, Let us make man in our image, after our likeness."**

Verse 27 continues:

> **"So God created man in his own image, in the image of God created He him; male and female created He them."**

(Emphases mine – both.)

Those two primary quotes from The Book of Genesis reveals, very clearly, what science and theology have not only **not understood**, but **missed completely**; *the second part of the answer.* And that is; that there are **two separate** "Creations of man".

Thus, in the above Chapter **1**, we have **both** *male* and *female* beings *created*, **but not out of the dust of the ground.**

Conversely, in Chapter **2**, (our first quote) we initially have only man being formed, **but from** the **dust of the ground.** Later, after a plea for company, the Bible narrative describes the first *woman* as being *fashioned* from a rib of the first *earth-man*. However earth-woman *arrived*, it nonetheless occurred **AFTER** the **First** *Creation* of "man" i.e., male and female.

Is there a contradiction here? No there is not! The Creation process described in Genesis has simply not been understood at all. This great degree of non-understanding has seeded the assumption for many that it all had to have taken place in the physical/material environment of the earth. Such an assumption would be perfectly valid if we were *only and solely* physical substance. Since we are not, then other *realities* obviously need to be considered.

So, from the standpoint of general Christian thinking, the acceptance of a single Creation-concept for the formation of man is regarded as the norm. Yet the orthodox Bible, from which the Christian Church takes its teachings and spiritual substance, *clearly states otherwise.* Anyone can pick up *almost* any Bible and find the same for themselves. So what should we make of this?

What we should not fear to undertake is a keenly searching examination of a possibly contentious religious issue whereby the deepest and most wonderful revelations are missed. Moreover, it should be exactly the role of the Churches in the first instance to **fearlessly** seek out the Truth, and to immediately discard any untruths discovered. That would be the right thing to do. With **bold courage and spiritual certainty**, that is *precisely* what we will do here!

At this point in our search for answers, it is timely to examine a Bible that is not accepted as possibly being "church-standard" but is, nevertheless, one that comes closest to providing what we now know to be the ***correct*** interpretations to the answers we seek. The following comparative passages are taken from: **"The Holy Bible in Modern English"**, by Ferrar Fenton. The Author of this remarkable **Work** heads the very first Chapter in The Book of Genesis with the words:

The First Creation of the Universe by God = Elohim.

This is clearly an exceptionally significant statement and a radical departure from orthodox thinking in that Fenton *identifies* a **First Creation**. In a comparison of the relevant Verses it is patently clear that the probability of a complete and fundamental misinterpretation of Genesis regarding the Creation of man has entrenched itself in our thinking for the last two thousand years.

Unfortunately, therefore, this critical error has brought Christendom, especially, to the point where most within would be too afraid now to even *question* this as just a *possibility*, let alone as clear truth. Such reluctance or fear, however, means that thereby we of the Western world – whose primary Religion and societal Law is both anchored in, and derived from, the Judeao/Christian ethos – have completely missed what is our rightful and ordained heritage: **The actual Truth of our Origins!**

So let us take the next steps boldly and examine, from Fenton's Bible, the *two* Creations of "man" in sequence. The *comparative* Verse 26 in Chapter **1** is preceded by the heading:

Creation of Man under the Shadow of God.

The Verse reads:

> "GOD then said, "Let Us make men under Our Shadow, as Our Representatives."

Verse 27 continues:

> "So GOD *created* men under HIS own Shadow, *creating* them in the Shadow of God, and constituting them *male* and *female*."

The actively-promoted notion by scholars and leaders of the three monotheistic religions that human beings – having transited through the earthly death-process – rise forthwith to 'heaven' to be in the very *immediate* presence of The Creator; in brutally-logical terms, *simply beggars belief.* How do they envision *their God?* Just as some kind of greater and stronger 'spiritual' or 'spiritualised' human being?

If we take the reality of just our Earth and Solar System as an example: It is powered by a small sun *so strong and bright* that its light will cause *immediate blindness* if we look directly at it. And that is a very small sun out of billions in just our galaxy alone. Our galaxy, as incomprehensibly immense as it is, is just *one* of billions in our universe in the Material Creation. [There is yet more, the extent and form of which we explain in Chapter 11.] That completely incomprehensible immensity, mind-blowing even in thought, is just the Material World only. By virtue of its materiality, therefore, it must inherently be; *and it is*: **The smallest and lowest part of the entire Creations!**

A life lived in a faith-belief doctrine that teaches *a close and convenient relationship* with a Power and Force that not only Created the vastness of the material Universes, but also the far greater Eternal Realms; at death is forced into the *shattering recognition* that we are not Created in the image of God at all. And, therefore,

31

cannot *ever be* in His immediate Presence. The Bible states it very clearly.

To reinforce this Truth, an appropriate quote from John 1, Verse 18 of, respectively; the King James Bible and Fenton's Bible notes:

"No man hath seen God at any time."

"No one has ever yet seen God;"

The demarcating barrier is there for all to read; which we elucidate in this Chapter.

0.2 The "Subsequent" Creation for Earth-man'

Now, very significantly, Verse 7 in Chapter 2 is also preceded by a relevant heading:

The Formation of Man from the Dust of the Ground by the Ever-living God.

Thus we have clear and unequivocal clarification of a huge and fundamental difference – the **first** as an ***immediate Creation***, **close to** The Creator. The *second* as simply a *forming*, **very far from** The Creator. The actual verse regarding the forming of man reads:

> "The EVER-LIVING GOD ***afterwards*** *formed* Man from
> the *dust of the ground*, and breathed into his nostrils the life
> of animals; **BUT MAN BECAME A LIFE-CONTAINING
> SOUL.**"

(Italics only; mine.)

The last phrase of the above sentence should be marked well because it provides ***the actual key*** to a true understanding of this whole question of our origins. Significantly, **it is printed in bold capitals in Fenton's Bible**. To state once more in reinforcement; the most significant aspect is the clear and unequivocal reference to the *first* happening; with man – both "male and female" – being *directly created*. That is in stark contrast to the second phase; with man being only ***formed*** – and from out of the "dust of the ground". The same *relative Scriptures*, moreover, appear in the same *respective places* in both the King James Bible and Fenton's, as it does in most Bibles.[6]

Thus far we have basically established that man is both a spiritual being and a physical one. And, moreover, further noted that only here on the earth can these "two parts to man" co-jointly exist as a single, completely whole and integrated entity. What has not been explored yet is *how* they are co-joined. Again The Bible holds the key. Since the complete process naturally implies that there had to be a beginning for man and his earthly home, this is so stated.

Let us return once more to "Sophie's World" and join her as she struggles to make sense of a concept that we may never be capable of grasping in its "living reality". That struggle to fully understand is due to the vast natural gulf existing between the Creative Power of all that exists, and us – the *lower **formed** entity* that is man. Such an idea is simply beyond our ability to ever comprehend.

> "They had learned at school that God created the world.
> Sophie tried to console herself with the thought that this

---

[6]It is probably pertinent, at this point, to mention a few facts about the author of The Bible from which the last passages were taken. In 1853 Ferrar Fenton resolved to study The Bible in its original languages and to re-translate it completely into English. Fifty years later he had accomplished his task of translating the complete Scriptures of the Old and New Testaments from the original Hebrew, Chaldee and Greek. Whilst the general thrust of the recognised story of Genesis is obviously known, there are seemingly small but extremely significant changes to some passages; changes which throw a whole new slant on some strongly entrenched beliefs. Throughout his work he explains translation errors, mainly in the Greek and Latin Versions, by showing where and how they occurred. Most importantly, however, the "small changes" he identifies allow for a vast expansion of perception regarding the clarification of the problem of interpreting the *time* taken for the Creation-process, not to mention the whole concept of our relationship with The Creator. Ferrar Fenton's intuitive insight into a more logical and correct explanation of the *seven days of Creation* resulting from his re-translation of The Bible may well signal a note of warning to purveyors of the status quo and to more recent Bible translators whose own efforts may have been clouded by religious preferences rather than a purely objective and logical analysis. For the purposes of this discussion, if – and I use the word *if* only in terms of this Essay being viewed as nothing more than an interesting hypothesis by some readers – *if* his re-translation of the Creation part of Genesis is correct, it might be wise to carefully consider whether this man's ordained Spiritual purpose was to help bring clarification to those Christian Churches and Bible translators who still hold to the literal view of seven earth days for the Creation process. Fenton's re-translation of Genesis in this case clearly offers a more stupendously correct interpretation, given its ability to accommodate both the viewpoints of scientific evolution and a Creation-process in a *logical* scenario and time-frame.

was probably the best solution to the whole problem. But then she started to think again. She could accept that God had created space, but what about God Himself? Had He created Himself out of nothing? Again there was something deep down inside her that protested. Even though God could create all kinds of things, He could hardly create Himself before He had a "Self" to create with. So there was only one possibility left: ***God had always existed***."

(p 8., Emphasis mine.)

Via this simple mechanism of logical elimination Sophie has hit on the only credible answer to the question – assuming, of course, that one accepts the belief of a Creator in the first place. Therefore, from the unequivocal acceptance of that premise – ***and that is the standpoint from which this 'essay' is written*** – in order to have Planes of Creation to "fill the void", including the material worlds of the observable universe, there had to be set in motion a "Creation-process" to bring this about.

So, through the stupendous and humanly-incomprehensible process of **Creation**, driven by the **Power** of **THE CREATOR** in the **DIVINE Ordination and COMMAND**: –

"LET THERE BE LIGHT!"

– the Creation-process of the forming of all the worlds in all the various planes began. As the lowest and therefore last *precipitation* of all the levels in Creation – and under the outworking of The Spiritual Laws of Creation, particularly *The Law of Spiritual Gravity* – the vastness of the *material* universes also came to be. Contained within just that lower immensity; our earthly home. Thus the *void* was filled.

Because of the sheer impossibility of ever being able to picture or understand what is for human beings an inconceivable process anyway – for what words in the many languages of the world could one possibly use to even attempt to describe it – what kind of earthly example could we employ to *try* to explain *how* the separation between the respective Planes of Creation occurred? If we are able to arrive at some small degree of comprehension of *at least the mechanics of the process* and thus why this demarcation was necessary, we can gift to ourselves a large measure of inner awareness of our *true place* in Creation. That kind of recognition should also help in the understanding of what our *Spiritual purpose and thus our actual reason for being* really is.

As a very basic and crude analogy, we can perhaps relate the main points of the Creation-process to that which occurs every hour of the day throughout the world in the numerous oil refineries of the petrochemical industry. This process is called distillation. It takes place in a "distillation column" whereby heat-generated crude petroleum vapour rises inside a tall metal column. At specific heights within the column, the vapour condenses to form various liquid petroleum products.

Each *different* distillate *will form itself* from the condensing vapour *at its appropriate condensing level as it cools*. This will be determined by the *weight and consistency* of the particular distillate being condensed *from the vapour at that level*. The distilled product at *each level* will thus precisely configure itself to *its own specific type of material substance*, and at its *specific temperature*.

I have highlighted particular terms in the previous paragraph in *bold italics* to help explain certain points that we will now outline. The explanations given here basically accord with all similar processes derived from the influence of gravity. Whilst this one example is earth-orientated i.e., operating from below upwards, the process of Creation naturally works from the Highest Heights *downwards* to the material worlds, but obviously still subject to The Law of Spiritual Gravity. Though we will never ever be able to even remotely understand this stupendous event – for even a combined distillation of the most ennobled aspects of all the world's languages could not even begin to offer a summary of words sufficient enough for the process – I believe we should nevertheless strive to achieve at least some small insight. Even a diagrammatical picture in the mind would be of value, for that is better than having no picture at all.

In the actual Creation process, then, we might perhaps envisage something approximating an unfathomably vast, white-hot mass of "downwards-moving" substance suffused with the stupendous Power of The Divine, vast enough to eventually form the incomprehensible immensity of the material universes –

– after the far greater Creation of the Higher Spheres: The First Creation!

And, as we must reinforce and strive to understand, all driven from the immediate proximity of The Light and Power of **The Creator** under the aegis of the "Creation-Words":

"LET THERE BE LIGHT!"

Because of the obviously immense power and pressure in closest proximity to The Creator, only the strongest and purest of beings could come to immediate consciousness there in the planes of their sphere of activity – i.e., *closest to GOD*. Thus, at heights we could never comprehend, and certainly never ever reach even in our *spiritual* form, occurred:

The Creation of Man in the First Creation.

Man *created* in the **Image of GOD**, (Gen. 1:27, Fenton.) – and **not** man *formed* later in the **second** Creation – from out of *the dust of the ground*. (Gen. 2:7, Fenton.)

Now, basically similar to the distillation process we examined earlier, the pressure of The Power of The Light drove the Creation-process to its completion. As each **different species within Creation** found its **appropriate level of consciousness**, as determined by its **weight and consistency** – i.e., its own **specific gravity** so to speak – so, too, could the planes for those inhabitants **form around them** in the **cooling-off process**.

This process was repeated all the way down to the material worlds. The governing factors which determined those levels of forming were the same as in our oil-refinery example – the *lighter and finer* in the **Higher Spheres**, the correspondingly *heavier and coarser* toward the **lower levels**. Thus each Realm formed itself at its appropriate place, corresponding to a level or plane whereby the *distance* from The Creative-Light permitted a *cooling off* and a condensing and thus an eventual *awakening to consciousness* there of that Realm's particular inhabitants.[7]

Only in the separating out and cooling off stage, roughly similar to the earthly process of *sedimentary deposition*, could worlds and landscapes form in which all the inhabitants of Creation would be able to fulfil their purpose, exactly as we must do here. For we should not suppose that only in the material sphere are there worlds of lands and rivers etc.. This difference is clearly alluded

---

[7]Each individual happening, each minute and incremental change in the cooling-off process clearly spans immense spaces and distances which, of course, we can never even begin to understand. The distances of interstellar space in the physical universes alone are simply too incomprehensible to grasp, never mind concepts of the vastly larger Realms of **The Spiritual** – or the even greater **Divine**.

to in Genesis, Chapter 1, where worlds, animals and fishes were *created*, before the **first** *Creation* of *male* and *female* in the **Image of God.**[8]

And only in Chapter 2, in an incomprehensible time-frame representing the aeons-long evolutionary process concerned with the *physical* world of Planet Earth, do we then find the animals, fishes and birds being "...formed from the dust of the ground". Earth-man, "formed" the same way – "from out of the dust of the ground" – thus enters his new world. It is a world of incredible diversity and pristine beauty.

> "And out of the ground the LORD God *formed* every beast of the field, and every fowl of the air: ..."

> (Genesis 2:19 King James. Italics mine.)

The same Scripture in Fenton's Bible states:

> "Therefore the EVER-LIVING GOD, who had *formed* out of the ground every animal of the field as well as every bird of the skies, took them to the man to see what he would name them. And whatever the man with the Living Soul called them, that was their name."

> (Italics mine.)

[The exact nature of the "forming" is of crucial import, for it reveals the *science* in the whole Creative process.]

The major difference between the lower and coarser physical worlds and the infinitely Higher and lighter Spiritual Realms is the *consistency* of the *substance* of which the *respective levels* are *composed.* Thus the First Creation is of *Spiritual* substance whilst the second, out of the dust of the ground for the earth and physical universes, is obviously material.

The key point to reinforce and understand here once again is that **each level represents a different consistency,** lightest and finest in the Highest Spheres – increasing incrementally with each subsequently formed level – until the heaviest and thus lowest in the material world. Such a far-reaching concept should not be all that difficult to understand. The statement of Jesus that: *"My*

---

[8]By 'The Spiritual' we do not mean 'near-earth' places generally associated with departed souls and occult or psychic activity, but a far greater spiritual reality. **"In my Father's house are many mansions."** [John, 14:2]

Kingdom is not of this world"; reveals the sure fact that His World [*"I came from The Father and I return to The Father"*] could not possibly be some kind of barren or filmy, amorphous expanse for He gives the strongest hint that His World is anchored in its infinitely more powerful and Eternal reality. Many of the great religions speak of an attainable Paradise if one lives one's life **based on The Laws of God** – [but not, however, according to the rules of Churches and Religions]. So every Realm above the material is thus progressively more paradisiacal.

Quite logically, therefore, the consistency and substance of the various Planes of Creation must necessarily be *the same* as that of *their inhabitants*. This *logical* progression implies that each Realm *must* also feel, and be, very firm and *real* for those who *reside there*; exactly as with humankind on earth. Any idea to the contrary is simply unrealistic. Nowhere in all the Spheres of Creation, therefore, do any of the inhabitants *float about aimlessly*, as is sometimes depicted in religious interpretations or films. Every inhabitant and every thing in the various planes is thus *anchored* into the *substance* of its particular "Realm" through the *consistency* of the level concerned.

The Eternal Laws operate throughout all of Creation, and the effects of The Law of Spiritual Gravity are felt in every sphere also. The Law of Movement, too, implies **activity** – everywhere. Thus we may note the perfect outworking of those Laws in the Creative process.[9]

Understandably, however, any sudden adjustment to this kind of conceptual perception may require a leap of quantum proportions. All we need do to achieve this, however, is to use the abilities of our inner Spiritual core, the real you and me. Abilities given to us precisely for this purpose; to understand our "Spiritual Origins" – explained in this 'essay'.

Now, what about those Spiritual origins; that of our higher Spiritual home? Because we are clearly only developed or *formed* beings – i.e., from *spiritual* **non-consciousness** to *personal* **self-consciousness** and not actually *created beings* as with man of "The First Creation"; as Fenton reveals – we did not therefore possess the inherent strength **to awaken to consciousness** *close*

---

[9]**The Spiritual Laws and Principles** to which we constantly refer may be read in their greater detail in the **Parent Work** – notated at the end of this Booklet – and in a stand-alone Sister Booklet: **"The Spiritual Laws of Creation *The Crucial Knowledge for Humankind"*.**

to GOD. Our level of *spiritual residence*, therefore, had to be *far lower* – in a kind of second-level Spiritual Realm. Yet even at that huge distance from *The Creative-Light Source*, we still did not possess sufficient strength to take on form and become conscious of self there either.

That state of *non-consciousness* therefore meant that we would require a home of transition – a material one – in which to *acquire* self-consciousness. Quite logically, any kind of material home could only be below that of "The Spiritual Realm". And only in that lower, material world would we be able to *develop to personal self-consciousness*.

Thus, in our non-conscious state at the very lowest levels of The Spiritual Realm – our true home – we, the future spiritual human inhabitants of the earth, awaited our time to *incarnate* in the material worlds far below.

> We awaited the completion of the *evolutionary develop-mental process* that would bring forth the *appropriate physical vessel* – that of *the primate* – via "*...the form-ing of man from out of the dust of the ground*". And into which *the immortal Spiritual aspect of man could first be placed* i.e., "*...the breathing into 'it' of the breath of life*".

Therein lies part of the understanding of the Creation of man. And therein, also, lies part of the reconciliation of the Creation-versus-evolution debate. That contention is, in reality, a totally unnecessary argument since there is no actual reason for this di-vision save that which the proponents of the two opposing views have "created" themselves.

The complete process still requires further clarification, how-ever. So apart from the earth being the place of transition for our awakening to self consciousness and spiritual awareness, was there a greater purpose for being permitted the opportunity for self-conscious life? Unequivocally yes! The material paradise of our earthly home was not only the place where we would develop to personal self-consciousness but, more importantly, *where we were to learn the truth of our Origins.* We were also tasked to protect and nurture the earth and its creatures given over to our stewardship as stated in Genesis and, having achieved "Spiritual Purity" through a voluntary adjustment to The Laws of Creation i.e., **The Rules**; we could then return, *ascend*, to our true home – The Spiritual Realm.

That particular Sphere is the promised *Kingdom of GOD* for human spirits. Thus, we are not even beings who stand close to The Creator, but just *developed ones* far from The Light. To reinforce this Truth, an appropriate quote from John 1, Verse 18 of the King James Bible notes;

"No man hath seen God at any time."

Fenton writes:

"No one has ever yet seen God;"

For the purposes of spiritual clarity, the understanding that the level of our Origin lies far below that of The Divine, the "Abode of The Creator", is absolutely imperative. This fact therefore forever precludes us from ever personally knowing the All-Powerful Creator we *too loosely and superficially* call **God**. The Eternal Laws unequivocally impose the completely natural barrier that a creature can only possess "actual knowledge" as an "inherent part of itself", **up to** its "source of origin". It is clearly not possible for any creature to *fully understand* levels beyond, or higher than, its own beginnings.

A simple but pertinent illustration is the difference in the level of intelligence and "awareness" between just animals and humans. How much *greater* must the difference naturally be between humans and **He** Who permitted us form and conscious life? To believe that we are at, or can attain to, the same degree of knowledge and power – as some scientists occasionally imply – is simply ludicrous and foolishly-arrogant. Just as ludicrous is the belief among some eastern religious groups that they will one day become "one with God".

Whilst we cannot "consciously know" more than that which our level of origin would permit, we can, however, *perceive* things from above such a level, as in the case of the Jewish people's intuitive recognition of the one invisible God when most of the rest of the world at the time were worshipping a variety of idols. We also possess the capacity to perhaps roughly *visualise* levels above our origins if we are given this information from One Whose Origin is from a higher level. The tidings and knowledge of the Higher Spheres given to us by Jesus is such an example. To believe, however, that we have, or can achieve, the ability to absolutely "know" in this way is incorrect. It is simply beyond the capabilities of even

40

our Spirit, whose actual home is from a far higher point in Creation than this lower-level material earth.

A simple test of this "truth" is to try to picture the concept of *infinity*, just as Sophie is attempting to do. To accept that there is a Creator logically means that there has never been a time when God did not exist. He has always existed. He will thus exist forever. We, on the other hand, need beginnings and ends as frames of reference to help in the understanding of everything connected with our existence and with time, and thus cannot even *begin* to grasp such a concept. The mind rebels and almost shuts down against such an alien thought because it has no affinity with such a far-reaching "idea". Only One who has no beginning and no end can logically "live" this kind of "infinite reality". For us, with our very limited perceptive ability, it is simply *an impossible thing* to grasp.[10]

For the moment, however, the problem of needing to completely reconcile the ongoing conflict between orthodox science and fundamentalist religion as to the origin of man has not yet been fully resolved. Some major points in previous paragraphs have offered many insights, but more explanations are needed. As previously stated, the real tragedy here is that this great difference of opinion exists only in the minds of the proponents of the respective opposing views, for it **cannot actually exist in reality.**

In other words, the pointless arguments that have marred this path since the initial stirrings of scientific thought brought the first rumblings of disquiet into the previously sacrosanct church view could not, quite obviously, have had *any bearing whatsoever* on the **actual** *forming* of the worlds in its stupendous scope and scale, however long ago it may have been. That reality will forever stand separate from all human opinion, as it obviously must.

Continuing to rigidly and stubbornly hold to a personal or professional viewpoint at all costs, and sometimes even against the quiet warning of the inner intuitive voice, is a sad reflection of much that is wrong with humankind. Yet for the sake of a clear, true picture of our origins and for peace of mind, resolving this totally unnecessary debate is imperative. However, this can only be achieved with a completely open, fearless and enquiring mind and, most importantly, without preconceptions.

---

[10]This concept may sit uncomfortably with some, though might serve to inculcate a more realistic attitude in our self-perceived relationship to the "Creator of all the Worlds".

The Bible once more offers the final resolution to this *apparent* quandary. The question here is one of interpretation or, more precisely perhaps, incorrect interpretation. The standard view of fundamentalist Christianity is that the "7 days of Creation" scenario – applied to the complete Creation process including the emergence of man – is non-negotiable, and probably because of the view that The Bible itself *seemingly* states that this is so. So strong has been this belief that it is now an entrenched and apparently immovable cornerstone for many. Rather than being a correct and thus sacrosanct anchor point for the Church, however, it is one that causes dissension and confusion.

Clarification of a previous key quote from Genesis should help to consign this division of opinion to its proper place in the sure relief of finally knowing the answer, thereby allowing the differences to be completely expunged. Hence, in the last phases of the great Creation, our world of the lower, material sphere could take on form too. As the last level of precipitation from out of the Creative process, this vast world of matter took billions of years to coalesce into roughly the form we know.[11]

That long, slow, evolutionary progression which subsequently emerged from the initial Creative-process eventually allowed for all material life forms to emerge, including "...the formation of man from out of the dust of the ground..." and the breathing into his nostrils; "...the life of animals." This is precisely what science has discovered. It was exactly that incredibly long evolutionary process which saw the emergence and preparation of the *physical vessel* – the development of the primate – that vessel which would ultimately house *Spiritual* man.

The preparation of the physical vessel for man is clearly revealed in Fenton's Bible where that particular happening is separated from the formative process by its denotation in his key capitalised phrase. The two primary events are revealed in this key Scripture:

> "The EVER-LIVING GOD afterwards formed Man from the dust of the ground, and breathed into his nostrils the life of animals, BUT MAN BECAME A LIFE-CONTAINING SOUL."

(Genesis 2:7)

---

[11]The vast world of matter referred to here is infinitely more than that which cosmologists believe they know. Its true nature and size is explained in Chapter 11 of the Parent Book.

Now, if this "sacred Scripture" is separated into two parts and simple logic applied to both, we discover a crucial point. From the King James Version, the first part reads:

"And the Lord God formed man from the dust of the ground..."

This small, seemingly innocuous, part-sentence actually holds one of the key components to resolving the Creation versus evolution debate between Christian fundamentalism and science. In it is revealed the science of Creation that quite clearly equates to that aspect of the overall ***Creation-process*** denoting the long evolutionary development of the physical vessel – the primate – ***formed from the dust of the ground*** to one day house "Spiritual man". The second part states:

"...and breathed into his nostrils the breath of life; and man became a living soul."

(Emphases mine. [Both quotes.])

Equally clearly, the "breath of life breathed into the nostrils" ***is the animating aspect for that physical vessel***. So simple yet so profound in concept, and so stupendous in scope and scale.

Perhaps many a reader may now recognise the correct picture with this explanation. For, as we state once more in reinforcement, the "...forming of man from the dust of the ground..." was simply the evolutionary process by which all creatures developed after the formation and cooling of the material earth: From the first minute microscopic life-forms out of the primordial soup, to the fishes, insects, plants, birds and great lizards, and thence from mammals to the first primates.

Thus did our **physical-form** ancestors slowly develop to their particular zenith – the refinement of form ordained for ***Spiritual-man.***

A marvellous and completely logical happening quite naturally divorced from any *fundamentalist* connotations when viewed correctly.

The associated aspect of the *time-frame* needed for evolutionary development and how that could possibly be reconciled with a complete Creation time of just 7 days as depicted in The Bible requires clarification too. The "7 days" account, though accepted by *well-meaning* Christians world-wide, is essentially, *and correctly,* rejected by the scientific community. Even the more recent new and supposedly *definitive* translations such as The Jerusalem Bible – compiled by committees of *learned theologians* – still *perversely persist* with a literal 7 days Creation-time. The scriptural quote: "...and a thousand years are as one day..." (2 Peter 3:8) scarcely suffices to place even the smallest dent in the time period required for **evolutionary development**, given that the dinosaurs alone reigned for some 180 million years.

Why, then, do such a large slice of 'Christian' humanity still persist with the absurd belief that just 7 earth-days – *equating to 7 earth-days of 24hr time* **as we experience it here on earth** – accounted for **'every' facet of the whole Creation-process**?

Do we really believe that just a few thousand years ago, dinosaurs marched into the Ark two by two, as some Creationists are 'desperately' striving to promote?

The true answer in this case is one of *incorrect interpretation* and non-understanding perhaps resulting from an incorrect translation of the original writings, or perhaps from simply accepting a symbolic *spiritual* term that was never meant to be so read. And then applying to it a literal, *earthly* point of view. It is, in effect then, **an incorrect "spiritual interpretation".**

WHAT, THEN, IS THE CORRECT INTERPRETATION?

The correct interpretation lies in recognising *the fundamental differences* between Chapter 1 and Chapter 2 of The Book of Genesis! The misinterpretation from Christian orthodoxy lies in attempting to ascribe two very different processes – one, **Creation**, the other, *primarily* **Evolution** – to a singular 7 earth-day Creation time-frame in Chapter 1, and attempting to also include in that time-frame the completely separate processes that Chapter 2 explains. That inexcusable error embraces the erroneous belief that the First Creation of Man – both male and female – refers to man on earth.

The scientific misinterpretation on the other hand lies in either completely disregarding the Creation aspect *as correctly outlined in The Bible*, and/or viewing Creation as a singular, material cosmological process out of which sprang the evolutionary developmental phase of the various earth creatures – including man – *through physical/genetic processes solely*. Fenton's Bible both delineates yet also harmonises the earth-science and Christian-fundamentalist points of view so completely that both are effectively *neutralised* in their *individual* positions, yet conjoined *perfectly* when brought together.

Special Note:

> *Perpetuated by the Christian Church, earthly science and committees of Ph.D and degree-toting Bible "scholars", this "triune" of earthly power and flawed education persistently clings to and continues to intellectually debate this foolish divide. Riding on the back of so-called "expert" translations and opinions, the main Bible Publishers continue to repro-*duce the same <u>appalling</u> and <u>inexcusable</u> error. *Despite this* dreadful <u>suppression</u> *of The Truth about Creation by the "educational elite" of the "Christian" part of global human-ity; we, with Fenton's essential contribution and guided in the same way that he surely was, will,* together, *reveal the actual processes that facilitated our entry into spiritual life on earth.*

So the "7 days of Creation" – **The First Creation** – described in Chapter 1 was therefore not immediately the Paradise of the human spirits, or the earth. It describes *actual* spiritual happenings at heights and distances *immeasurable* and thus *inconceivable* to earthly humanity. We should therefore not become confused with the term, earth, used in the account of Creation in Chapter 1. That word *does not* refer to any kind of "local" association with our planet. It must be understood as a *"concept of Creation"* which applies to "dry land".

> "And God called the dry land 'earth': and the gathering of the various waters He called 'seas'. "

(Genesis 1:10)

In The First Creation, therefore, there are also mountains, forests, meadows, seas, animals and men – as we have previously and strongly noted – but of inconceivable beauty and perfection as prototypes for all subsequent Spiritual Creations, all of which could only come into being *after* The First Creation. Thus, in Genesis 1:11-13 (emphasis and parentheses mine), it is stated:

> "Let the earth (the dry land) produce seed-bearing vegetation, as well as fruit trees according to their several species, capable of reproduction upon the Earth;" and that was done. The Earth (the dry land) produced the seed-bearing herbage according to every species, as well as the different species of reproductive fruit trees; and GOD saw that they were good. This was the close and dawn of the third *age*.

Now, what is this new and very different word – *age* – describing the Creation-process? A quick comparison of our two main reference Bibles reveals a vastly different contrasting picture with exactly that one small word making all the difference. The King James Version of Genesis 1:1 reads:

> "**In the beginning** God created the heaven and the earth."

This Bible, as with most others, goes on to state in Chapter 1 Verse 5:

> "And God called the light Day, and the darkness he called Night. And the evening and the morning were the *first DAY*."

And so on to Verse 31:

> "And God saw everything that he had made, and, behold, *it* was very good. And the evening and morning were the *sixth day*."

And thus to Chapter 2, Verses 1 and 2, which state:

> "Thus the heavens and the earth were finished, and all the host of them. And on the *seventh DAY* God ended his work which he had made; and he rested on the *seventh DAY* from all his work which he had made."

Surely we are able to acknowledge that this word, 'day', as used here, is nothing more than a symbolic term for a particular "time-period"; one which was never ever intended to be taken literally. The example from *Verse 4* of the *same Book* quantifies this premise because it states:

"These *are* the **generations** of the heavens and of the earth when they were **created**, in *'the DAY'* that the LORD God made the earth and the heavens."

(All emphases mine.)

As previously stated, the literal acceptance of the word, **day**, in the singular for the complete Creation of both the heavens and the earth – coupled with the unequivocal and clear reference to the word **generations** in the same creative phase – quite obviously calls into question any literal acceptance of "7 earth days" for any kind of complete Creative-process. Clearly there is an immediate and obvious contradiction here.

In stark contrast, Ferrar Fenton's *correctly* re-translated Bible Verse of Genesis 1:1 reads very differently:

"By Periods GOD created that which produced the Solar Systems; then that which produced the Earth."[12]

It is crucial to understand here that a vastly *different* conception of time must inherently exist in spheres that are obviously *non-material*. The diurnal rhythm we experience here on earth simply cannot apply to such levels, for the passing of time can only be experienced in *material* planes. In Realms that are Eternal, time, as we *believe* we know it, simply does not exist. Time, therefore, does not **pass** there. We on material and finite earth, however, **experience its passing** every second of our earthly existence.

So in comparison with Verse 5, Chapter 1 of the King James Version, Fenton states:

"And to the Light GOD gave the name of day, and to the Darkness He gave the name of Night. This was the close and the dawn of the **first AGE**."

And so on to verse 31:

"And God gazed upon all that He had made, and it was very beautiful. Thus the close came, and the dawn came of the **sixth AGE**."

And Chapter 2, Verses 1 and 2 comparatively state:

---

[12] Fenton's translation of the word, "Periods", equates literally to, "By Headships". He writes: "It is curious that all translators from the Septuagint have rendered this word (- as -) B'reshitii, into the singular, although it is plural in the Hebrew. So I render it accurately. – F. F."

47

"Thus the whole Host of the Heavens as well as the Earth were completed. And GOD rested at the **seventh AGE** from all the works which He had made."

In similar vein, Verse 4 reads:

"These were the productions for the Heavens and the Earth during their Creation at the **'PERIOD of their organization'** by the LORD GOD of both the Earth and the Heavens."

(Emphases mine.)

And **only after** that *direct Creation* phase did God **then** subsequently *form* earthman – via a long evolutionary process – from "...out of the dust of the ground...", along with the animals and birds etc.. This scenario fits perfectly with the scientific view of a very long developmental phase for the earth after its birth before even the ancestors of the very earliest primates could emerge.

It is patently clear, therefore, that the difference between an "earth-day" and an **"age"** is immense indeed. Simple logic should suggest that the designation of an **age** to denote a **phase of Creation** utterly stupendous and incomprehensible to human thinking in its scope and scale makes eminently more sense than the literal acceptance of an earthly 24 hour time-period that Christian fundamentalism has perhaps interpreted as meaning a Genesis *"Creation day"*.

And neither should that correct view clash with any religious interpretation regarding the awesome greatness of the Creation-process by The Creator Himself. Religious fear and blind faith do not supply answers. They only serve to perpetuate spiritually-wrong concepts with the resultant effect of producing adherents too afraid to think for themselves. Perhaps the explanations outlined here may induce the tentative religious reader to become *less afraid* to question worrying uncertainties.

After all, how often do we hear the common phrase or variations of: *"It wasn't like that in my day!"* Generally used to compare an earlier time-period in a particular life to that of the present, the connotations here are surely obvious to all. Such references do not refer to a single day.

0.3 The 'Real' Human/Chimp Split: The 'How' and The 'Why'

Even though the previous explanations offer solutions to some contentious questions, there is still the need to provide further clarification to particular points for our complete elucidation. In one area at least, however, our discussions thus far should offer the clear recognition that we, *as the complete entity man*, are **not descended from the primates**. The physical body which we inhabit today is a refinement of form that developed from the very first primates during the long and natural evolutionary process that eventually led to upright man.

Therefore, let us re-state once more that this "body-form" is nothing more than *a physical vessel only*. It is, however, one we absolutely need in order to live in the material environment of the earth. The animating power that separates itself out from the physical shell at earthly death, that some in the scientific community refer to as "the ghost in the machine"; *that* is the real you and me – our individual Spiritual core whose home and Origin is *not of the earth*.

The above explanations may also answer the question as to what happened to the many varied species of primates that disappeared virtually overnight, and why the search for a so-called missing link is a difficult one. With the entry of the human/spiritual aspect into the most advanced species, all others *striving to develop to that same point* – but who *could not provide* the necessary *strength of attraction* for the *Spiritual part of man then reaching the earth plane* – simply died out. As a natural consequence of insufficient development and the inability to compete with this newly arrived "Spiritual force" in the shape of early Spiritual man, all other primate groups that *could have* developed to become *that particular vessel* were rendered superfluous.

Now, whilst we have a resolution to both parts of this religious/scientific debate, *how* the human/spiritual aspect actually entered the appropriately developed primate species or group requires further clarification. For only then could the true "human being" now multiply within its own new species and thus populate the earth and develop into the world's peoples.

We have established the fact that the human being is both spiritual and physical and possesses the attribute of *free will*. It is also clear that without the *Spirit* as the animating force within the human body, the physical vessel has no life of its own i.e., it is naturally and necessarily dead. In this particular state, it follows the decree of The Laws of Nature and decomposes. It can thus be deduced that all other *mobile* life-forms of the earth, such as birds, insects and fishes etc., must similarly have an inner animating core in order to have life. Their inner core is, however, not spirit, but **soul**, from which is derived the discernment of **instinct**.

As with man on earth, animals, too, have an original home above the material world from where they draw their life-force. However, because they do not possess the attribute of *free will* – which is synonymous with *personal spiritual responsibility* – their *place of origin* is **below** that of the human spirit. That plane of origin is designated as the Animistic, or Elemental Realm; which is that of the Nature Beings and the Elemental Forces of Nature. Whilst these other *mobile* creatures of the earth do not possess *free will*, they do have the driving power of the *instinct*. This inherent attribute allows all such creatures to develop their ordained place and purpose too.

> It is therefore *absolutely crucial* to recognise *and understand* that **up to** the incarnation or arrival of Spiritual-man onto the earth, *animal-man* [earthly primate already here] **possessed _soul_ as his life-force**. And, in "evolutionary-ordination", further developed the *vessel* through which the entry of the **human/spiritual** onto the earth plane could occur.

To this end, over the many millions of years of evolutionary development of the earth's creatures to the perfecting of the form of *animal man*, the human spirit (us), slumbering in a state of non-consciousness in the lower levels of the *Spiritual Realm*, obeyed an *unconscious, inherent urge* to *develop* to *personal self-consciousness*. Since this could not take place there, and as a result of that "inner petition", the non-conscious spirit-seed – in terms of its individual journey – was *driven or expelled* downwards towards the material worlds from *out of* The Spiritual Realm. Thus, our true home.[13]

---

[13]Quite logically, the *actual* paradisiacal **Garden of Eden** must *naturally exist* in a higher Spiritual Realm. And *that* is **the reality**. The notion that human beings on earth are driven out of such a place to somehow *still be on the earth* is strange indeed, and completely incorrect. A similarly-incorrect idea of earth-based human thinking – the nature of **The Holy Grail** – also *obsesses* over a solely *earth-based location and connection* for its ostensible place. See Sister Booklet: – [**JESUS! His Birth Death and Resurrection; A Revisionist Analysis of the "Sacrosanct" Christian Viewpoint**] – for full clarification.

This is the mechanism whereby we "petition" to be born; precisely to develop to self awareness of who, what, and why we are!

Now, as the human "spirit-seed"[14] traverses the intermediate planes below The Spiritual Realm all the way down to the Material Plane, it is compelled to accept a covering or cloak of the same *consistency* as that of the plane through which it is journeying. In the great immensity of what we refer to or understand as being "the beyond", there are "many mansions" too. The "lower mansions" of Creation are the Realms or levels we must traverse on our journey to earth in order to gain conscious life.

Thus, the closer the *now-stirring spirit-seed* gets to earth, the *heavier* its outer coverings become. We may refer to them as the bodies or coverings that pertain to, and are of, the particular *consistency* of the *different Realms* traversed. Through this process, we become more than just an entity comprising a Spiritual core [the real us] and a physical body. In reality we possess a number of "cloaks" or coverings, each one telescoped into the other, so to speak, with each one having a different consistency exactly *commensurate* with the *sphere* or *plane* from which it is *derived.*

This *collective* covering is the *soul-body*, distinct from the *inner Spiritual core*. The final cloak, of course, is that of the physical body, taken on when the *Spirit*, with its enveloping "soul-coverings", *enters, incarnates* into the growing foetus of a pregnant woman under the perfect outworking of **The Law of Spiritual Attraction of Similar Species**. This outer physical body becomes the *earthly vessel* for that *complete soul body* and its *enveloped spirit.*

The processes that determine whether a male or female is born in this first incarnation will have been decided by the journeying spirit's *intuitive inclination* as it travels downwards toward the material world and the earth. From an initial *non-conscious* state, it gradually begins to slowly *awaken* the closer it gets to the world of the physical. During its transition downwards, it begins to sense the life-stream currents of the various levels through which it passes.

As it descends further, it becomes more and more firm *in its inclination towards the experiences it wishes to make its own.* With this increasing certainty comes the firm decision to choose either an *active* and, therefore, *male incarnation*; or a

---

[14]Note the **Parable: "A sower went forth to sow."**

51

passive and, therefore, ***female one***. [Passive in this context does not mean weakly- or meekly- submissive.] Thus, the *nature of the activity it chooses* **determines its body form.**

This explains the *mechanism* of *personal* choice that *determines* whether the entity's final form will be male or female. From that point onwards, in its successive incarnations, it *can* change its outer forms depending, of course, on whether it changes its *activity*. Thus it is possible for a spirit to incarnate in alternate male and female bodies, which enables it to experience and develop both the male and female abilities inherent in the Spirit. However, it can also happen that such a situation generates for that individual, spiritual and emotional uncertainty as to its true ordained place.

> *In this explanation lies the key to the many emotional and/or psychological problems of humankind with regard to sexual orientation.*[15]

Now the stage is almost set for the entry of the human/spiritual into the world of matter, therein to determine his future spiritual outcome. The earth, up to this point inhabited by prolific numbers of varied creatures, did not yet know Spiritual man, nor the effect that this particular creature would exert on their natural, pristine world. Unknowingly awaiting the advent of this new stranger were the most highly developed species of primate quietly furthering their role as the chosen vessels through which this event would be fulfilled.

With the unnecessary conflict between science and religion in ***one*** aspect of the "Creation versus Evolution" debate incontrovertibly resolved, let us complete the process by examining the ***second*** part of our wondrous yet contentious Genesis Verse to determine the final happening. This is tightly bound to our old friend, Verse 7, Chapter 2 in Genesis of the King James Version of The Bible. As previously stated, two distinctly separate parts to this Verse can be readily identified. Quoted in full once more, it reads:

> **"And the Lord God formed man of the dust of the ground, and breathed into his nostrils the breath of life; and man became a living soul."**

What we are now concerned with are the words:

---

[15]Everything has its answer – its origin and end – under the *immutable and inviolable outworking* of **The Spiritual Laws.**

"...and breathed into his nostrils the breath of life..."

As previously refuted, this surely cannot be taken to mean that God Himself descended to earth to literally blow His breath into a model of a "mud-man". Though brief and simple, this part of the Scripture clearly depicts something else. It depicts the **animating** of the future **physical vessels** for humankind, facilitated through **"...the forming of man from out of the dust of the ground..."** That "forming" occurred quite naturally and logically during the aeons-long *evolutionary process*. Through those vessels – *the especially prepared species of primate* – man would eventually exert his new and far-reaching ***Spiritual influence*** in the material worlds at the ordained time for this to occur.

Now, in order to answer, from The Bible, the question of *how* the entry of the human/spiritual aspect into that species of highly developed primates prepared for its reception was effected i.e., those possessing a soul as their animating life-force – *animal man)* – we need only re-visit Verse 7, Chapter 2, but *this time* using Ferrar Fenton's translation of Genesis to find the *true* connection. Fenton offers that *actual revelation.* It is one, moreover, which gives a far clearer understanding than we have thus far found in any other "scriptural writings". So, once again we read:

> "The EVER- LIVING GOD afterwards formed Man from the dust of the ground, and breathed into his nostrils the **life of animals**; – **BUT MAN BECAME A LIFE-CONTAINING SOUL.**"

> (Bold emphasis mine.)

The *difference* in wording here from commonly accepted Scripture represents a huge leap forward in our knowledge of the understanding of Creation and may well reflect Ferrar Fenton's *intuitive* grasp of the true happening to enable him to detail it *more* precisely, but without *actual* confirmed knowledge.[16]

---

[16] Fortunately that new knowledge is available today, hence the unequivocal nature of this Essay. Fenton's clearer "spiritual insight" enabling him to clarify the terrible error/distortion about Creation – so long accepted without question – is revealed toward the end of the "Explanatory Note" of his remarkable work and sublime 'Calling'. It now finds voice here too. To those few of his assailants who 'sneered' that his work was 'not a translation but a mere paraphrase', he writes: "The remark shows that they do not know the difference between one and the other, or a perusal of my rendering of the Hebrew of the *two first chapters of Genesis*, and my note thereon, ...would show to them the purely philological basis of my translation." (Italics mine.)

The fact that the evolutionary development of the *primates* necessarily required the symbolic "*...breathing into the nostrils the life of animals...*", simply reflects the creative and actual necessity for **all mobile creatures of the earth to have an inner animating core**, their actual **life-force**. Thus that which is referred to here as "*...the life of animals...*", is 'their' *inner life-aspect*, which all animals must possess in order to live.

It is that of soul.

As we have learned, man *also* possesses a soul – *but not as his innermost 'animating component'*. The primary life-force of man – the actual person – is **spirit**.

His **spirit** – *he* – is *enveloped* by *his soul*, the *multi-layered* outer cloak or *covering*. In necessary reinforcement:

Spirit is who and what man is; *when all else is stripped away*.

So the reference to the word "soul" pertaining to man in this particular and *correct translation* of the Biblical account must be understood in its *true meaning*, since there is clearly a fundamental difference between the life-force inherent "***...in the man breathed into...***" and **what he then became** as a result of this **breathed-into process**. Moreover, there is a stated delineation with the use of the conjunction – **BUT!**

This can be further extended to logically mean, "**...but *nevertheless became*...**" Fenton must have understood the crucial difference here because this key phrase is capitalised in *his* Bible, i.e.: **"BUT MAN BECAME A LIFE-CONTAINING SOUL."**

We may perhaps better understand this most crucial difference which separates man from animal by examining Fenton's footnoted reference 1 immediately after the word, 'animals', in the Scripture: i,e., "...and breathed into his nostrils the life of **animals1**." Fenton's footnote states:

"1 Or reflective or intellectual life. See Cor. Ch ii.12. Ch iii.3."

We know, of course, that animals are not reflective or intellectual in the human sense, but that is not what is meant here.

So if we now look at Corinthians 2:12, we really do begin to understand what Fenton must surely have *intuitively* understood:

"And we have *not* received the *spirit of the world*, but the **Spirit proceeding from God**; so that we can *distinguish* the gifts God has granted to us."

"...for you are animals still. For when there is *rage and strife and dissensions among you*, are you not rather like *animals*, than conducting yourselves like men."

(Corinthians 3:3, All emphases mine.)

Note:

If there is still uncertainty regarding these explanations in the mind of you, the reader; here, now, is the final note on this *most crucial and necessary of revelations for **all** of earthly humanity*!

In reinforcement *once more*: We now know that the breathing "...into his nostrils the **life of animals**..." represented the entry – *at the very beginning* – of the *inner, animating life-force* into the very earliest *physical-form ancestors* of primates at **their forming** from *'out of the dust of the ground'*, out of which would *eventually develop* the most *highly evolved species* to exist on earth. And from *that* 'broad evolutionary spread' the *further development* of the *single primate branch* which would *subsequently produce* the "physical-form vessels" *human beings would require* for life on earth.

So the great and fundamental **unbridgeable difference** between primates and humans is the fact that even though we also needed to use the basic physical form of the most *highly evolved* of the primate species for *our human* earth-life activity too –

– we could not do so with just the 'inner, animating life-force' of the primate.

For we are not animal, we are human.

And we therefore could not just simply **evolve into humans** from the primate.

Why not? Because, *once again*, that which gives human beings life *is not* the same inner life-force as that which the lower-level animal primates possess as *their* life-power. Thus, the capitalised reference to **"LIFE-CONTAINING SOUL"** reveals exactly that fact.

That which is *contained in, and enveloped by, 'the soul'* in this crucial revelation is ***The Spirit – THE LIFE** – the innermost animating core of the 'human being'*.

Therefore, the key capitalised phrase: **"BUT MAN BECAME A LIFE-CONTAINING SOUL."** reveals that the *quantum leap* from primate to human was *not one given life by human genetics* through simple 'brain development' by normal but slow 'evolutionary processes'. It was a quantum leap –

– precisely because a new force and power was required for humans to be humans.

Thus completely distinct and very far removed from the *animal* primate. That is why the 99 percent of similar DNA in both *monkey and man* concerns the respective *physical bodies only*, and thus why **they both rot away at death.**

We are not our physical body!

We can conclude, therefore, that this inherently logical process that Ferrar Fenton reveals is that which actually facilitated the entry of *Spiritual man* onto the earth.

It thus depicts the amazing event of the *incarnation*, the *entry*, of the **human/spiritual** from out of his former *non-conscious* state in the *non-material* Spiritual Realm, into the most highly developed **species of primate** then existing on earth. Only thereby could man *develop* to become fully conscious of self.

'Spiritual' man – 'human' – now stands on earth!

The reader who supports Bible Scripture literally and solely may now begin to see that there is no conflict after all between the Creation account and evolutionary development since both, in their *co-joined natural perfection*, could only have issued from the hand of The Creator Himself in any case. Utilising Hegel's 'dialectic process' and Fenton's correct Bible translation, we have succeeded in bringing together the appropriate connecting threads from both the scientific and religious disciplines to offer the reader *definitive clarification* of the Creation-process.

So there stands our Spiritual Origin; that of all humankind. From that stupendous happening, Spiritual man, in his material home prepared over millions of years, began his new journey of development towards personal self-consciousness and knowledge. This journey, however, would span a certain, precisely ordained period of allotted time requiring accountability at the end of it. For with his request for conscious life came the responsibility of *correct stewardship* of his material home; the earth and all its creatures, as was once **Commanded**!

The division and separating out into the races, peoples and languages of the earth still lay before him. History records that stewardship as one of mostly degradation, destruction, blood and war, with few intervals of true peace, grace and nobleness. Now, however, the time of accountability has arrived. We stand in the midst of the reciprocal effect of our bad stewardship as The Spiritual Laws set about the grim task of "balancing the books". That necessary "auditing process", which we now begin to observe with some considerable degree of alarm, is effected via the increasing outworking and activity of the unassailable power of the "Elemental Forces of Nature"!

The clarification herein of humankind's origin and entry into the material world of the earth as home should provide meaning and insight for many readers. With clarification should also come quiet confidence and peace of mind as to who and what one really is. Here, we outlined and explained the basic happening of the coming-into-being of Creation, and of man. We accompanied him on his journey downwards, from a non-conscious spirit seed in The Spiritual Realm to incarnation and conscious personal responsibility on the earth, and thence to the employment of his free will to determine his spiritual future under the outworking of The Eternal Laws.

Naturally, we should expect that there is a lawful process by which he leaves this earth too. Whilst considerable detail is given to this process in the **Parent Work** [**Chapter 9 "The First Death"**] it is appropriate to conclude here with a short paragraph to very briefly outline the reverse process of his [our] return journey, thus completing our cycle of *spiritual* life in this segment. [The reader should note, however, that this particular outline only explains the return ascent *if* The Spiritual Laws have been heeded. The process for a human spirit who **chooses** a path *opposed* to those Laws is, unfortunately for that human spirit, a very different one.]

So, upon earthly death and *release from the physical shell,* **and**

57

providing he has earned the right to do so, he begins his *return journey upwards* to his true Spiritual home. As he passes *into each correspondingly lighter sphere* – **the same that he traversed on his journey downwards** – *the heavier covering of the previous lower level is automatically discarded in accordance with its corresponding weight and that particular level's corresponding density.* In this manner his ascent continues until, finally, he stands at the threshold of The Spiritual Realm from whence he originated.

This time, however, not as a non-conscious, unknowing spirit-seed, but as a fully conscious, purified, spiritual being – the true Spiritual Man. Here, he sheds the last cloak and is drawn across that boundary into his Eternal home, radiant in his Spiritual Purity having **earned** "...the crown of eternal life". Thus is fulfilled the invitation of Jesus who stated:

> "In the home of my Father there are many abodes. If it were not so, I would have told you: because I am going to prepare a place for you."

> (John 14:2)

In order to earn this right, however, the next scriptural quote from Matthew 5:26 should also be accepted:

> "I tell you indeed, that you will not depart until you have repaid the very last farthing."

Therefore, to this end under The Laws of Creation i.e., spiritual/foundational science, we read;

> "You, however, should be perfect, as your Father in heaven is perfect."

> (Matthew 5:48, Fenton all.)

Given the possibly contentious yet potentially mind-extending implications of this essay, detailing the main points of the Biblical sequence of Creation here at the conclusion may provide simplified clarification via a greatly condensed overview.

0.4 Chronology of the Creation Process

0.4.1 Key Points

The Utterance of the stupendous Creation-Words – "**LET THERE BE LIGHT!**" – thus resulting in: **The First Creation – (The Spiritual Realms.)**

1. The Creation of the Heavens and the Earth of **The First Creation** – "By **Periods** God created *that which produced* the Solar Systems: then that which produced the Earth."

    (Genesis 1:1, Fenton, Emphasis mine.)

2. The **Creation** of day and night (in the Heavens.)

3. The division of the waters which were **under** the expanse (firmament) from the waters which were **above** the firmament (expanse.) The firmament/expanse then named the Heavens.

4. The commanding of the waters **below** the Heavens to be collected in one place, and for dry land to appear.

5. The **Creation** of flora.

6. The Creator sets two great lights which divide day and night for earth.

7. The **Creation** of fish and bird life.

8. The **Creation** of animal life.

9. Then, the great **Creation** of man **in His Image** – both male and female – and the Blessing to rule over all flora and fauna.

10. The **completion** of the Creations at the end of the **sixth Age**. The Creator rests at the **seventh Age** and blesses and hallows the seventh **day**.

Note Scripture: Genesis 2:1 (Fenton) "Thus the whole Host of the Heavens (as well as the Earth were completed.)" This is the completion of The First Creation (i.e. Spiritual Realms.)

(Parentheses mine.)

And only then:

> **The Creation of the Worlds of Matter,** including our universes, solar systems and earth – **as planned by its Creator.**

11. After the completion of **The First Creation (The Spiritual Realms)** including all that was then **created** (as described in Genesis 1:1-3), **The Creation of the Worlds of Matter** through a long process of evolution leading to the forming by God of earth-man from out of **the dust of the ground** who, following a suitable time of evolution, became the first human being – **the man with the 'Living Soul'.**

> (Genesis 2:19. All emphases mine.)

12. Earth-man gives name to every creature – **formed** from out of the **dust of the ground also.**

13. Even though God had **already created** "man" in His Own Image (both male and female) in **The First Creation,** and had subsequently **formed earth-man** from out of **the dust of the ground,** there was still no earth-woman. (Biblical tradition states that she was constructed from a rib of the man.)

Now, whether this sequence is viewed literally, symbolically, pseudo-scientifically or any other way, there are clear pointers illustrating a number of very different and very distinct happenings that occurred. This is clearly contrary to the one single sequence that the main churches generally believe and accept as having ostensibly **created** man/earth-man. And, moreover, him only. Fenton unequivocally delineates these separate, stupendous events in clear sub-titles.

1. **The First Creation of the Universe by God = Elohim.**

2. **Creation of Man under the Shadow of God.**

3. **The formation of Man from the Dust of the Ground by the Ever-living God.**

Therefore, if we take careful note of points 3, 4, 9, 11 and 13, and similarly note the above sub-titles 1, 2 and 3 from Fenton's translation, a vastly different and more stupendous picture arises than the present, general belief of an aspiritual humanity somehow being in close proximity to a Power that we cannot even begin to comprehend. Points 3 and 4 *on their own* here strongly indicate two different places very far apart: one above the Heavens and one below the Heavens. The Creator of all the Worlds is clearly far further from us than we might want to believe. (In any case recognition of that fact arrives to all not too long after earthly death.)

The sequence we have outlined here (from The Bible) concurs with many of the scientific findings of anthropology and astronomy. Taken in concert, both views actually trace a path of evolutionary development that is consistent with rational logic and, moreover, encompasses and co-joins both the religious and scientific points of view. More importantly, however, this more logical sequence places man (us) in his correct place very far from a Creative Force that inherently possesses the Power to Create all that we know – and all that we do not – simply by an Act of Will. In short, by the Power of Creative thought. That is a power and an ability utterly incomprehensible to us. Such stupendous Power, moreover, will forever be beyond our extremely limited comprehension.

However, instead of that particular recognition being somehow problematical for us, we should be grateful for the fact that the incomprehensible immensity of the Creations graciously provides all that we will ever need for life here – **and life eternal**. And should thus actually offer the largest possible measure of *inner peace and spiritual security possible*. That is provided, of course, that we voluntarily choose to – **obey the Rules!**

So therewith are the long-contested arguments of **Creation versus Evolution** – Christian fundamentalism versus intellectual science:

Herein perfectly reconciled and harmonised!

Bibliography

1. The deeper knowledge in this Booklet is derived from the Work: '*IN THE LIGHT OF TRUTH*', The Grail Message by Abd-ru-shin. 3 Volume Edition, Stiftung Gralsbotschaft Publishing Co., Stuttgart, Germany.

2. *The Gathering Apocalypse and World Judgement*, Crystal Publishing, 2005, New Zealand.

3. *The Holy Bible in Modern English*, Ferrar Fenton, Destiny Publishers, Massachusetts U.S.A. 1966 Edition.

4. *The Holy Bible, Authorised (King James) Version*, Eyre and Spottiswoode (Publishers) Ltd., Great Britain.

5. *The Gospel of the Essenes*, The original Hebrew and Aramaic texts translated and edited by Edmund Bordeaux Szekely, Revised Edition, London, C. W. Daniel, 1976.

6. *The Jerusalem Bible, Reader's Edition*, First published 1968, Darton, Longman and Todd Ltd., London.

7. *The Christian and Reincarnation*, Stephen Lampe, Millennium Press (UK) 1990.

8. *Building Future Societies*, Stephen Lampe, Millennium Press (UK) 1994.

9. *The Concise Oxford Dictionary of Proverbs*, 1983 Edition, Oxford University Press, First Printing 1982, USA, New England Journal of Medicine.

10. *Ideas and Opinions*, Albert Einstein, Bonanza Books, New York, 1954.

11. *Philosophy History and Problems*, Third Edition, Samuel Enoch Stumpf, McGraw-Hill, USA 1983.

12. *The Nature Of The Gods*, Cicero, Penguin Classics, 1972 Edition, Translation by Horace C P Macgregor, Printed by Richard Clay (S E Asia) Pte. Ltd. Reprinted 1978, 1984.

13. *Sophie's World*, Jostein Gaardner, Phoenix House, Great Britain, 1996 Edition.

14. *Autobiography of a Yogi*, Paramahansa Yogananda, First published 1946, Random House.

15. *Man's Eternal Quest*, Paramahansa Yogananda, Collected Talks and Essays on Realizing God in Daily Life, Volume 1.

0.5   The Parent Book:

Formerly:

"The Gathering Apocalypse and World Judgement; What It Brings – Even Now – And Why" [See Back Cover.]

Available in the **U.S.A** at
http://www.crystalbooks.org

In **New Zealand** at:
http://www.publishme.co.nz

Now:

BIBLE "MYSTERIES" EXPLAINED
[Revised Second Edition]
Understanding "Global Societal Collapse" from The "Science" in The Bible;
What Every Scientist, Bible Scholar and Ordinary Man Needs to Know!

> The **Revised Second Edition** of this book is more comprehensive in that it now explains How and Why the 2008 global economic collapse occurred, but also when the seeds that wrought the How and Why were sown, and by whom. [Chapter 3: **The Spiritual Laws: The Necessary Knowledge**
> 3.3.3 "Ten Men Will Take Counsel And It Will Come To Nought."
> 3.3.4 The Interlinked Global Monetary System "Reaping The Whirlwind." A Brief History Lesson.]

> Additional information about the events surrounding the last day of Jesus's life, from His arrest in the Garden of Gethsemane to His murder at Golgotha, is now included.
> The interesting question of the "Seven Churches in Asia-Minor" from The Book Of Revelation is examined more critically. Necessarily using the discoveries and mathematics of present-day cosmology, the revealing conclusion of the true meaning perfectly resonates with the intuitive perception of the great mathematician, astronomer, theologian and scientist, Sir Isaac Newton.

This book, the result of many years of inner seeking and empirical research, offers *serious* seekers of the Truth a comprehensive understanding of the origin, meaning and purpose of human life; material and spiritual.

Beginning with **The Crucial Imperatives: Nine key points** that *must* be taken into consideration if logical and reasoned answers to humankind's Whence, Whither and Why is *ever* to be understood; the book takes the reader step by step through an understanding of man's **Spiritual Origins, The Spiritual Laws of Creation**, the difference between **The First Death** and **The Second Death; Elemental Lore** [of Nature]; **Jesus! His Birth, Death and Resurrection** [a revisionist analysis]; before examining the truly 'mind-expanding' meaning of **"The 7 Churches in Asia"** from **The Book of Revelation.**

The key knowledge helps explain *why* there actually are **Two Sons of God**– final Chapter. It is key precisely because all other knowledge stems from that reality.

On reading the Work, the genuine seeker will clearly see that a conditioning process, set in place by religious authorities from the outset, over millennia has wrought appalling suffering through their inexcusable distortions of the Teachings of **The Truth** that once issued pristine and sublime from the Pure Holiness of its Bringer: **Jesus, The Son Of God!**

Now, because of those distortions, humankind is as a rudderless wreck on an increasingly stormy sea. Our many and increasing problems were not brought upon us by any kind of arbitrary randomness, but through *our constant and stubborn refusal to live according to the very Laws of Life which **alone** guarantee knowledge, peace and harmony.*

At the same time, however, – and precisely through the knowledge of those Laws – the way is shown in *how* we can *change* global societies *for the better.* Quite logically, if we continue down our present path for much longer *without such change*, the immutable outworking of **The Law** *will simply bring to an end* all that which *human thought and endeavour* had sought to establish and/or erect *in place of* the immutable and inviolable aegis of: **The One Law!**

CREATION-LAW!

The Parent Work explains the How, the What and the Why!

Available in **U.S.A.**: **http://www.crystalbooks.org**

Or – **N.Z.**:**http://www.publishme.co.nz**

Table of Contents

4 Elemental Lore Of Nature

5 JESUS: His Birth, Death and Resurrection

6 Stigmata

12 THE TWO SONS OF GOD!

THE BOOKLET SERIES

*   *   *   *   *

THE TWO SONS OF GOD

The Son of Man and The Son of God
What The Bible Really Says

*   *   *   *   *

JESUS!:
His Birth, Death and
Resurrection

A Revisionist Analysis of the "Sacrosanct"
Christian Viewpoint

*   *   *   *   *

THE SPIRITUAL LAWS OF CREATION

The Crucial Knowledge for Humankind

*   *   *   *   *

WHITHER COMETH HUMANKIND?
(The Origins of Man) *Genesis and Science Agree*

*   *   *   *   *

THE "7 CHURCHES" Of THE
"REVELATION"

What the "Hubble" Will Never See
Sir Isaac Newton's "Plan of The World"

*   *   *   *   *

www.ingramcontent.com/pod-product-compliance
Lightning Source LLC
Chambersburg PA
CBHW071847020426
42331CB00007B/1897